T0140199

Data Acquisition Systems

Maurizio Di Paolo Emilio

Data Acquisition Systems

From Fundamentals to Applied Design

 Springer

Maurizio Di Paolo Emilio
EDM Engineering
Pescara, Italy

ISBN 978-1-4899-8741-9   ISBN 978-1-4614-4214-1 (eBook)
DOI 10.1007/978-1-4614-4214-1
Springer New York Heidelberg Dordrecht London

Printed on acid-free paper

Springer is part of Springer Science+Business Media (www.springer.com)

To my best grandparents: Emilio and Zita
To my women: Julia and Elisa

Intellectuals solve problems; geniuses prevent them. [A. Einstein]

Foreword

On a recent visit my 11-year-old nephew—who is a very logical and empirical young man—said to me: "When I grow up I am going to be an engineer, just like you and Daddy!" To me that sounded like a reasonable wish, nevertheless I still asked him why so. "I want to build Formula 1 racing cars like Daddy and then write about them like you." It is good to hear today's children still find motivation and inspiration in the modern world, where we take everything for granted. Many away from engineering might say: "It works, so I don't need to know how or why!". But, at the base of this "everything" is engineering and the engineers who translate wishes, wants, and needs into components and systems that end up in a seamless, interconnected network of devices. It is safe to say that it is indeed the engineers— with their questioning minds, pragmatic approach and relentless tinkering—who make most of the world move forward. With a PhD in physics, Dr. Maurizio di Paolo Emilio is an engineer and one with deep understanding of the data acquisition world. Maurizio worked at the National Laboratory at Gran Sasso in Italy and collaborated with the University of L'Aquila on a fellowship project focusing on the software and hardware development of data acquisition systems. With an extensive involvement in the design of various data acquisition systems, Maurizio is well positioned to outline the requirements and internal workings of such projects. Data acquisition is a necessity; we rely on measurements to analyze and create various systems, but in order to build the right measurement system for an application, a good understanding of the considerations associated with every part of that data acquisition system is a must. Maurizio's new book is a comprehensive breakdown of the elements and steps involved in data acquisition, from the fundamentals through to the hardware and software needed, to final design examples. With the dedication of engineers such as Maurizio, who help educate and train the new generation of engineers, which the world relies on and desperately needs, we look forward to seeing what the young children of today, like my nephew, might bring to engineering—and the world!—tomorrow.

London   Svetlana Josifovska
                                                                          Editor, Electronics World

Preface

Data acquisition system (DAQ) is used to acquire information from some physical phenomena. The main process is to sample the signals that convert the analog value (electrical signal) of the sensor to a digital one and be manipulated by a computer. What's the main components of the data acquisition system? The main electronic systems can be one of the following members:

- Sensors: to convert the physical phenomena in electrical signal
- Analog-to-digital converter: to convert analog signal to digital signal
- Multiplexer and amplifier: to switch and amplify the input's analog signals with an analog–digital converter in digital form
- Display/computer: to visualize/manage the data

The complexity of new physics experiments requires more complex DAQ with the following characteristics:

- Capable of managing large amounts of data
- High-speed connection for DAQ
- Digital recording
- Full reconfigure possibility

Signals that are hard to characterize and analyze with a real-time display are evaluated in terms of the following parameters:

- High frequency
- Large dynamic range
- Gradual changes
- Sudden, unpredictable events (Fig. 1)

The goal of this book is to give the theory and practical information about the design of data acquisition systems; it helps to solve problems about the functional design hardware (and software). DAQ software allows us to communicate with and control the card; analyze and present the data. As for the software, here we will provide the general and basic concepts; the language programming details

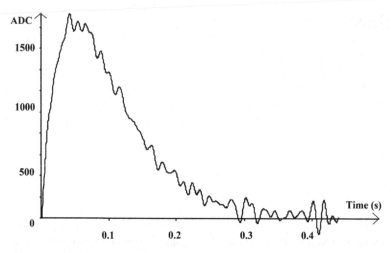

Fig. 1 Transient single shot event

for further elaboration are excluded. Each complex concept has been made in an easy-to-understand way which makes them readily usable. Handling the parameters is important for many industrial applications; main idea of data acquisition systems projects is to develop an application which can handle the sensitive parameters, such as temperature, humidity, and so on. Many of them use also GSM/UMTS/HSPDA technology or similar, though GSM provides voice and video calling facilities. Moreover, a possible high-speed DAQ will be described and we will analyze the theoretical aspects and possible configurations management by a computer through the main communication bus: USB and Wireless. In particular, it will be further discussed the design techniques introduced with functional design.

In the last year, some aspects are emerged to define the impacts of the evolution of the data acquisition system. For example, USB bus improves both of the ease of use and flexibility. Wireless has also a consolidation technology and standardization. For high-speed system, IEEE 802.11 has seen its adoption in mobile applications. Tablet computers will also be part of the future of DAQ systems, software has been and will continue to be the driver for many new application and technology changes for the future using more efficiently graphics user interface. Future of the data acquisition systems is based on the use of onboard field-programmable gate arrays (FPGAs). The flexibility makes them ideal not only for custom data acquisition requirements but also for embedded applications and the development and testing of custom digital device.

Pescara, Italy   Maurizio Di Paolo Emilio

Acknowledgments

I would like to express my gratitude to all those who gave me the possibility to complete this book. In particular, I want to thank Maxim Integrated Corp. for the kind support with very useful data sheet/application notes to the compilation of the book. Moreover, many thanks to Svetlana Josivofska, Editor of Electronics World, for the contribution of the foreword and Springer, in particular Charles B. Glaser Executive Editor—Electrical Engineering, for the publication of this book.

Contents

Chapter 1
Introduction

Abstract The main action of data acquisition systems is the sampling signals that measure real world physical conditions (voltage, current) and converting the resulting samples into digital values that can be handled, for example, by a computer. This chapter describes the fundamental concepts of data acquisition systems; in particular sensors, transducers, communication cabling, and system configurations.

1.1 Fundamentals of Data Acquisition Systems

Data acquisition (DAQ) systems are the main instruments used in laboratory research from scientists and engineers; in particular, for test and measurement, automation, and so on. Typically, DAQ systems are general-purpose DAQ instruments that are well suited for measuring voltage or current signals. However, many sensors and transducers output signals must be conditioned before that a board can acquire and transform in digital the signal. The basic elements of DAQ are shown in Fig. 1.1 and are:

- Sensors and transducers
- Field wiring
- Signal conditioning
- DAQ hardware
- DAQ software
- PC (with operating system)

Transducers can be used to detect a wide range of different physical phenomena such as movement, electrical signals, radiant energy, and thermal, magnetic, or mechanical energy. They are used to convert one kind of energy into another kind. The type of input or output of the transducer used depends on the type of signal detected or process controlled; in other ways, we can define a transducer as a device that converts one physical phenomena into another one. Devices with input function are called sensors because they detected a physical event that changes according to

M. Di Paolo Emilio, *Data Acquisition Systems: From Fundamentals to Applied Design*,
DOI 10.1007/978-1-4614-4214-1_1, © Springer Science+Business Media New York 2013

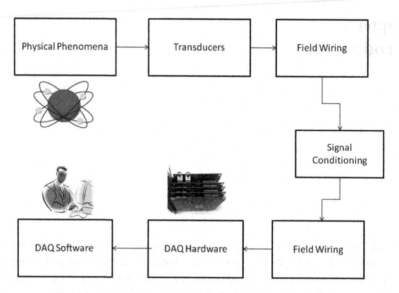

Fig. 1.1 Functional diagram of a PC-based data acquisition system

Table 1.1 Common transducers

Quantity being measured	Input device (sensor)	Output device (actuator)
Light level	Photodiode photo-transistor solar cell	Lamps–LED–fibre optics
Temperature	Thermistor–thermocouple	Heater–fan
Force/pressure	Pressure switch	Electromagnetic vibration
Position	Potentiometer–encoder	Motor
Speed	Tacho-generator	AC/DC motors
Sound	Carbon microphone	Buzzer–loudspeaker

some events as for example heat or force. Instead, device with output function are called actuators and are used in control system to monitor and compare the value of external devices. Sensors and transducers belong to category of transducers.

There are many different types of transducers; each transducer has input and output characteristics and the choice depends on the goal of your system: for example from type of signal that must be detected and the control system used to manage it (see Table 1.1).

Sensors produce in output a voltage or current signal according to the variation of physical phenomena that are measuring. There are two types of sensors: active and passive. Active sensors require external power supply to work; instead, passive sensors generate a signal in output without external power supply. Signal conditioning consists in manage an analog signal in order that it meets the requirements of the next electronics system for additional processing. Generally, in various applications of control system there is a sensing stage (e.g., a sensor), conditioning stage, and a processing stage. The conditioning stage can be built, for example, using operational amplifier to amplify the signal and, moreover, can include the filtering, converting, range matching, isolation, and any other processes required to make sensor output suitable for processing stage. The processing stage manages the signal conditioned in other stages such as analog-to-digital converter, micro controller, and so on [1].

1.2 Sensors and Transducers

Transducers and sensors are used to convert a physical phenomena into an electrical signal (voltage or current) that will be then converted into a digital signal used [2] for the next stage such as a computer, digital system, or memory board.

1.2.1 Temperature Sensors

Several techniques for detection of temperature are currently used. The most common of these are resistance temperature detectors (RTDs), thermocouples, thermistors, and sensor ICs. The choice of one for your application can depend on some factors such as: required temperature range, linearity, accuracy, cost, and features. RTDs are more commonly known; they are built using several different materials for the sensing element, for example the Platinum. Platinum is used for different reasons: high temperature rating, very stable, and very repeatable. Other materials used for RTD sensors are nickel and copper.

Thermocouple is composed of two different metals that have a common contact point where it is produced a voltage (some mV) proportional to the variation of the temperature. Thermistors are generally composed of semiconductor materials. There are thermistors with positive and negative temperature coefficient. The thermistors with negative temperature coefficient are used to monitor low temperature of the order of 10 K [2–4]. The temperature coefficient is defined from the following equation (1.1).

$$\alpha(t) = \frac{1}{R(T)} \frac{dR}{dT} \tag{1.1}$$

In general, a linear curve is used for working only over a small temperature range. For accurate temperature measurements, it is necessary to use the Steinhart–Hart equation (see (1.2)):

$$\frac{1}{T} = a + b * \ln(R) + c * \ln^3(R) \tag{1.2}$$

where a, b, and c are parameters. The solution of (1.2) can be written as (1.3):

$$R = e^{(x - \frac{y}{2})^{\frac{1}{3}} - (x + \frac{y}{2})^{\frac{1}{3}}} \tag{1.3}$$

where

$$x = \sqrt{\left(\frac{b}{3c}\right)^3 + \frac{y^2}{4}} \qquad (1.4)$$

and

$$y = \frac{a - \frac{1}{T}}{c} \qquad (1.5)$$

Typical values of the resistance of $3,000\,\Omega$ at room temperature ($25\,^\circ C$) are the following:

- $a = 1.40 \times 10^{-3}$
- $b = 2.37 \times 10^{-4}$
- $c = 9.90 \times 10^{-8}$

1.2.2 Magnetic Field Sensors

Magnetic sensors convert magnetic energy into electrical signals for processing by electronic system. Magnetic sensors are designed to respond to a wide range of magnetic field; they are mainly used in different applications, in particular in automotive systems for the sensing of position, distance, and speed. For example, the position of the car seats and seat belts for air-bag control or wheel speed detection for the anti-lock braking system (ABS). Magnetic sensors work according to the Hall Effect (see Fig. 1.2): the production of potential difference (Hall Voltage) across a conductor where a perpendicular magnetic field is applied.

The output voltage, called the Hall voltage, (V_H) of the basic Hall Element is directly proportional to the magnetic field (B) passing through the semiconductor material:

$$V_H = R_H * \left(\frac{I}{t} * B\right) \qquad (1.6)$$

where R_H is the Hall Effect coefficient, I is the current flow through the sensor in Ampere, and t is the thickness of the sensor in mm. Most commercial Hall effect devices are manufactured with built-in DC amplifiers, voltage regulators to improve the sensors sensitivity, and the range of output voltage that is quite small, only few microvolts [2].

1.2.3 Potentiometers

A potentiometer is an electromechanical device that contains a movable wiper arm with the goal of maintaining electrical contact with a resistive surface; the wiper

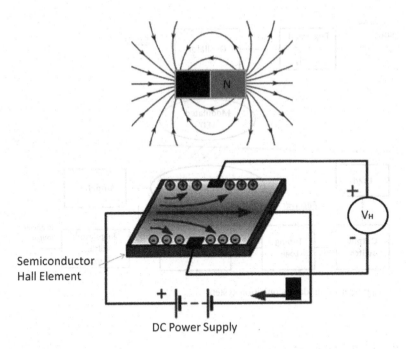

Semiconductor
Hall Element

DC Power Supply

Fig. 1.2 Hall effect sensor

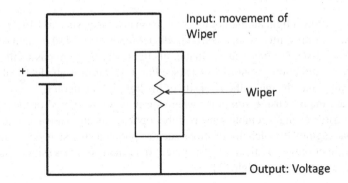

Input: movement of
Wiper

Wiper

Output: Voltage

Fig. 1.3 Potentiometers

is coupled mechanically to a movable linkage. It gives a voltage signal by divider circuit when voltage is applied across the entire resistance within the potentiometer. See Fig. 1.3. A variable potential difference can then be produced at a central wiper arm relative to one of the resistor as the wiper is moved. The wiper is usually made of a material such as beryllium [1].

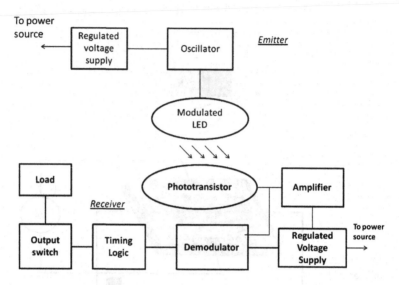

Fig. 1.4 Light detection: data acquisition system

1.2.4 Light Detection

Light sensors detect light emitted or given off from an object, such as LED, reflected from surfaces, transmitted from electronics device, and so on. LED or light emitting diode is a solid-state semiconductor that emits light when current passes through it in the forward direction. A photoelectric (see Fig. 1.4) sensor is an electrical device that respond to the change in the intensity of the light falling upon it.

There are many sensing situations where space is too restricted or the environment too hostile even for remote sensors. Fiber optics is an alternative technology in sensor "packaging" for such applications such as photoelectric sensing technology. Moreover, fiber optics are flexible, transparent fiber made of glass (silica). It works as a waveguide to transmit light [1].

1.3 DAQ Hardware

DAQ system can be defined as a set of electronic systems, which any of the following functions:

1. The input: processing and conversion to digital format using ADCs. The data are then transferred to a computer for display, storage, and analysis.
2. The processing: conversion to analog format, using DACs. The analog control signals are used for controlling a system or process.

3. The input of digital signals, which contain information from a system or process.
4. The output of digital control signals.

In general DAQ hardware is the interfaces between the analog signal and a PC. It could be in the form of modules that can be connected to the computer via serial and USB port for example, or cards connected to slot in the mother board: for example PCI or PCI express [1].

1.4 DAQ Software

DAQ software is the main component of the DAQ system and is needed in order for the DAQ hardware to work with a PC. DAQ software (see Fig. 1.5) can be written in a variety of languages (e.g., C language) and can be written for the particular application in design. Alternatively there are a number of proprietary DAQ software packages that are available and these can be utilized (see National Instrument with Labview). Usually, DAQ software is composed of a text-based user interface (TUI) consisting in an ASCII configuration file and a graphic user interface (GUI) available by means, for example, of any Web browser. Both interfaces permit DAQ management and customization without the need of recompiling the

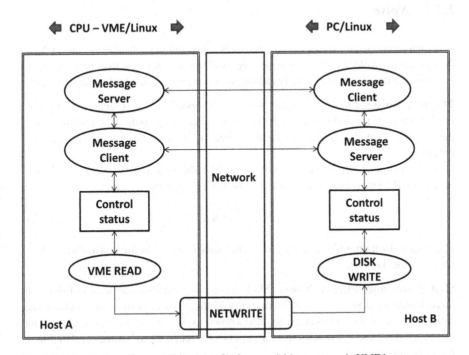

Fig. 1.5 Example for software architecture of a data acquisition system via VME bus

sources, thus granting full acquisition control also to inexperienced programmers. The configuration file is written in a high-level language (meta language) and is easily modified by the operator. The GUI works at a higher level with respect to the ASCII configuration file and helps the operator in compiling the configuration file and in controlling the acquisition. The use of the Web interface does not require any knowledge of the configuration file syntax and avoids "grammatical" errors. It is up to the operator to choose the TUI or the GUI when modifying the DAQ setup [1].

1.5 Communications Cabling

Field wiring is the physical connection from the transducers/sensors to the DAQ hardware. When the signal conditioning and/or DAQ hardware is remotely located from the PC, then it is necessary to use field wiring that provides the physical link. In this case, it is very important to estimate the effects of the external noise, especially in industrial environments. In the next paragraph it provides an estimation of this noise [1].

1.5.1 Noise

One characteristic of all electronic circuits is represented by Noise: it is a random fluctuation in an electrical signal generated by electronic devices. In communication systems, the noise is an undesired random disturbance of a useful information signal.

Thermal Noise: Johnson–Nyquist noise or Thermal Noise is generated by the random thermal motion of electrons. Thermal noise is approximately a white noise: the amplitude of the signal can be described by a Gaussian probability density function.

The root mean square (RMS) voltage due to thermal noise v_n, generated in a resistance R (ohms) over bandwidth Δf (hertz), is given by

$$v_n = \sqrt{4k_{\mathrm{B}}TR\Delta f} \qquad (1.7)$$

where k_{B} is Boltzmann's constant (joules per kelvin) and T is the resistor's absolute temperature (kelvin).

Shot Noise and Flicker Noise: Shot noise in electronic devices consists of unavoidable random statistical fluctuations of the electric current in an electrical conductor. Moreover, Flicker noise, also known as $1/f$ noise, occurs in almost all electronic devices, and results from a variety of effects, though always related to a direct current [1, 5, 6].

1.6 Parameters of a DAQ System

To properly design a DAQ system, we must know some important parameter. The goal of this section is to describe major system parameters for a better design [5].

1.6.1 Accuracy and Precision

In the fields of science, the accuracy of a measurement system is the degree of closeness of measurements of a quantity. The precision of a measurement system, instead, is called reproducibility or repeatability of measurements. Relative accuracy is a measure that indicates the capability of the DAQ systems to correct output codes according to its full-scale range.

1.6.2 Noise

Each measurement generates noise as combination of more signals. It is an interference between two terminals. One factor, common-mode noise, indicates the interferences that appear on both measurements inputs. The majority of common-mode interference is attributable to 50 Hz (or 60 Hz) power frequency.

1.6.3 Settling Time

The settling time of an electronic device is the time elapsed from the application of an ideal step input to the time at which the value output has entered and remained within a specified error range. Parameters that can describe settling time are the following: propagation delay and time required to obtain output value (Fig. 1.6) [5].

1.6.4 Acquisition Time

Acquisition Time is a feature of the DAQ systems that indicates the presence of an analog-to-digital converter. It defines the time for going from one situation to a new one according to the system accuracy [5].

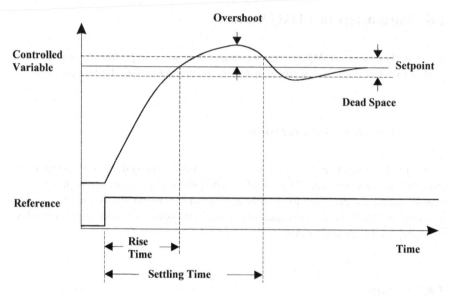

Fig. 1.6 Settling time

1.6.5   DC Input Characteristics

It indicates the value of offset voltages, offset currents, and bias current of a electronic devices.

References

1. Park J, Mackqy S (2003) Practical data acquisition for instrumentation and system control. Elsevier, Oxford
2. Lacanette K (1997) Temperature Sensor Handbook – Application Note National Semiconductor Corporation
3. National Instruments (2002) Data Acquisition Fundamentals, Application Note 007, National Instruments Corporation
4. National Instruments (1997) Signal conditioning fundamentals for PC-based data acquisition systems, Application Note 048, National Instruments Corporation
5. Taylor J (1986) Computer-based data acquisition system. Instrument Society of America, Research Triangle Park
6. Wikipedia http://en.wikipedia.org/wiki/Noise_%28electronics%29 – Wikimedia Foundation, Inc.

Chapter 2
Data Acquisition Systems: Hardware

Abstract This chapter describes different hardware aspects of the data acquisition systems, in particular the main components such as signal conditioning, analog-to-digital converter, and so on. Moreover, different techniques of interface between analog and digital signals to plug-in boards will be discussed.

2.1 Introduction

Capabilities and the accuracy of data acquisition (DAQ) systems can be founded from analog input specifications. Basic specifications is as follows: number of channels, sampling rate, resolution, and input range. The number of analog channel inputs will be indicated for both single-ended and differential inputs.

The most common way used to transmit electrical signals over wires is the form single-ended. In general, one wire carries a voltage (the signal), while the other wire is connected to a reference voltage (ground). Instead, differential signal is a method of transmission of electrical signals with two complementary signals sent on a couple of wires.

The main advantage of single-ended over differential signaling is that fewer wires are needed to transmit multiple signals. A disadvantage of single-ended signaling is that the return currents for all the signals can be shared with the same conductor, and this can sometimes cause interference ("crosstalk") between the signals.

Differential signaling technique can be used for the following applications [5, 6]:

- Balanced audio
- Digital signaling
- RS 422, RS 485, PCI express and USB
- Printed circuit board
- Connectors (BNC)

M. Di Paolo Emilio, *Data Acquisition Systems: From Fundamentals to Applied Design*, DOI 10.1007/978-1-4614-4214-1_2, © Springer Science+Business Media New York 2013

2.2 Plug-in DAQ Systems

Many DAQ systems, known as plug-in boards, are used in scientific application to acquire data and transfer it directly to computer memory. Transference of data can be done by parallel port, serial port, USB port, Ethernet port, and so on [1, 3].

Typically, DAQ plug-in boards (Fig. 2.1) are general-purpose DAQ instruments that are well suited for measuring voltage or current signals or resistance that can include some form of signal conditioning [1].

Usually, application in real time of high frequency analog signals needs a high speed DAQ with a dedicated plug-in processor such as a digital signal processing (DSP) board.

The main components of DAQ system is the analog input (A/D) boards; it converts (Figs. 2.2 and 2.3) analog voltages from external signal sources (sensor, see Chap. 1) into a digital signal, which can be read by the host computer. Moreover, the functional diagram of a typical DAQ system can be described of the following main components:

- Input multiplexer
- Input signal amplifier
- Sample and hold circuit

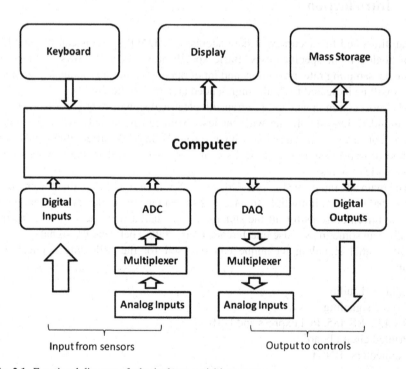

Fig. 2.1 Functional diagram of a basic data acquisition system

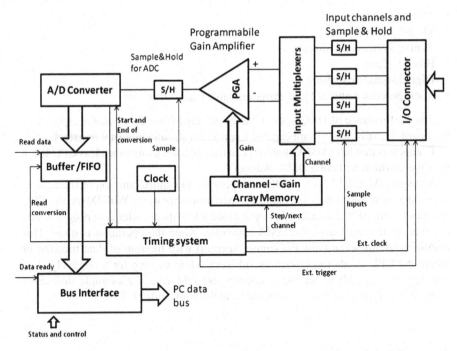

Fig. 2.2 Functional diagram of a generic A/D board

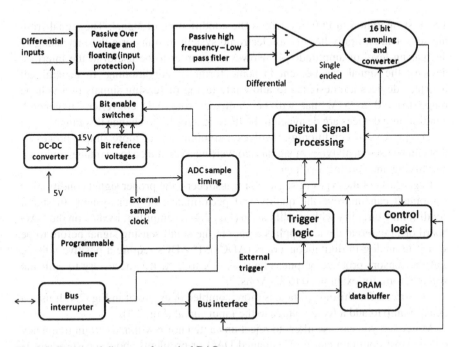

Fig. 2.3 Functional diagram of a typical DAQ system

- A/D converter
- Memory (DMA)
- Timing system and filtering
- Bus interface
- Digital signal processing
- Microprocessor and/or Field-Programmable Gate Array

The performance of DAQ system can be affected from data transfer capacity of the board. Usually, all PCs are capable to program I/O and interrupt transfers.

Computers that have DMA channels can transfer data using much less CPU laden than those which without a DMA channel.

Without DMA, CPU is typically occupied for the entire duration of the read or write operation, and is thus unavailable to perform other work. With DMA, the CPU initiates the transfer, executes other operations while the transfer is in progress, and receives an interrupt from the DMA controller when the operation is done. This feature is useful when the CPU cannot keep up with the rate of data transfer, or where the CPU needs to perform useful work while waiting for a relatively slow I/O data transfer. Many hardware systems use DMA, as for example disk drive controllers, graphics cards, network cards, and sound cards [1, 5, 6].

2.3   Signal Conditioning

The first operation in DAQ system is the conditioning of signal. The signal from the sensor must firstly be sent to electronic system to transform the increasing or decreasing level of amplitude [4]. However, in order to be useful to the interface devices, the signal can be sent in some forms of conditioning. In general, all interface devices are designed to allow interfacing of sensing signals in a voltage range from 0 to 5 V that will be digitized from A/D converter. In general, conditioning devices are designed to be flexible in order to change this range.

The process of conditioning involves a combination of more simple processes that can be used in all types of signals: converting a resistance to a voltage, dividing, amplifying and shifting a voltage.

Regardless of the types of sensors or transducers, the proper signal conditioning equipment can improve the quality and performance of your system. In signal conditioning circuitry with amplification (Fig. 2.4) an amplifier, located on the DAQ board or in external device, applies a gain to the small sensing signal before to be sent it in analog-to-digital converters (ADCs). If ADCs require a positive voltage and the sensors produce output voltage as $-X$ to $+X$, it is necessary to shift the voltage from, for example, 0 to 2X Volts.

The circuit for shifting voltage is more complex than circuit in Fig. 2.4. It uses a dual op-amp to add a fixed voltage to the input signal (Fig. 2.5).

Moreover, you can use filters to reject unwanted noise within a certain frequency range. Most common cause of damaged DAQ or problems about measurement is

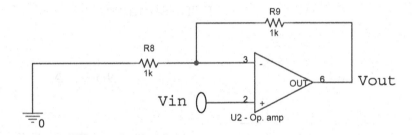

Fig. 2.4 Signal conditioning: amplification

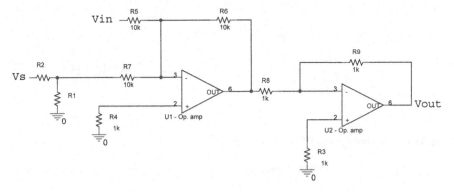

Fig. 2.5 Signal conditioning: shifting voltage

improper grounding of the DAQ system. In order to avoid that, it is necessary to isolate signal conditioners and prevent most of these problems [1, 2, 6, 7].

2.3.1 Example of Design of a Signal Conditioning Circuit

First of all, we must determine input signal range [4]. In this example of design, RTD sensor (resistive Temperature Detector) in Wheatstone bridge configuration will be used as our input signal. The goal of this circuit is to measure temperature in range 0–70°C, and our input will be the potential difference between Node A and Node B.

Resistance of RTD varies from 100Ω to 127Ω which represents 0 to 70°C temperature range. Voltage difference varies from 0 V to 91.94 mV, in steps of 3.74 mV per 1Ω increase (Fig. 2.6).

Next step is to design a circuit that can detect output potential difference between node A and node B. INA333 (instrumentation amplifier) will be implemented for such purpose. INA333 is a type of differential amplifier that does not require input impedance matching through two input buffers. From DC analysis the maximum voltage difference between two nodes is 91.94 mV, so in order to increase resolution

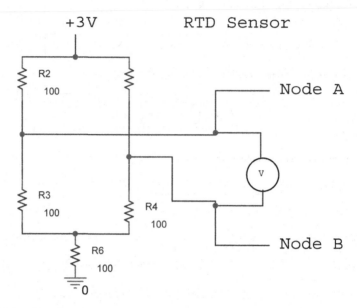

Fig. 2.6 Signal conditioning: PT1206

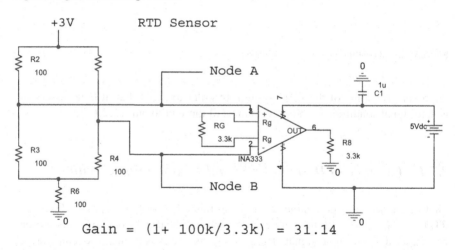

Fig. 2.7 Signal conditioning: amplification

and signal-to-noise-ratio (SNR), we design the amplification circuit with gain of approximately 30; in this way it matches (for example) with reference voltage of ADC in use (Fig. 2.7). After amplification stage of signal conditioning, signal must be filtered and optimized. The design of RC filter is determined by following four parameters of ADC: acquisition time, sampling ADC input capacitance, time constant multiplier, and full-scale input voltage range (Fig. 2.8).

Capacitance is set to be at least bigger than 20 times of ADC input capacitance.

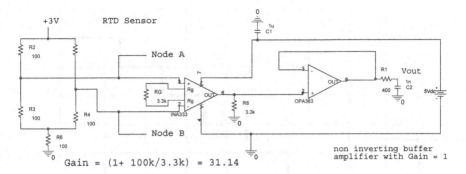

Fig. 2.8 Signal conditioning: filtering

Using a capacitor with capacitance of 1 nF with other parameter, equivalent external resistance has to be at least larger than 398 ohms, which appropriate buffer amplifier needs to be chosen. OPA363 from TI can be ideal as buffer amplifier due to the low common mode rejection ratio (CMRR) and optimized to be used as a driver for ADC input capacitance (Fig. 2.7) [3, 4, 6, 7].

2.4 Converters A/D

The connection of digital circuit to sensing device can be done only if the sensors are inherently digital themselves. However, when analog signals are involved in the project, the interface becomes much more complex. In this case, it needs a way to translate analog signals into digital form: an ADC; digital-to-analog converter or DAC performs the opposite operation.

Typically, an ADC (Figs. 2.9 and 2.10) is an electronic device that converts an input analog voltage or current to a digital number proportional to the magnitude of the voltage or current. However, some non-electronic or only partially electronic devices, such as rotary encoders, can also be considered ADCs. The number of output bits from an ADC doesn't fully specify its behavior. Real A/D converters can differ from ideal behavior in many ways. While static imperfections, such as gain and offset, are easy to quantify, the success of many signal-processing applications depends on the dynamic behavior of the A/D converter. Ultimately, the application determines the requirements, and A/D converter resolution may not be either necessary or sufficient to specify the required performance. In many cases, the quality of the A/D converter must be tested for the specific application. The wide variety of ADC applications leads to a large number of figures of merit for specifying performance. These figures of merit include accuracy, resolution, dynamic range, offset, gain, differential nonlinearity, integral nonlinearity, signal-to-noise ratio, signal-to-noise-and-distortion ratio, effective number of bits, spurious-free dynamic range, intermodulation distortion, total harmonic distortion, effective resolution bandwidth, full-power bandwidth, full-linear bandwidth, aperture delay,

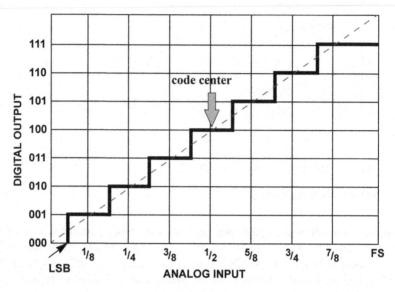

Fig. 2.9 Ideal characteristic of analog to digital converter

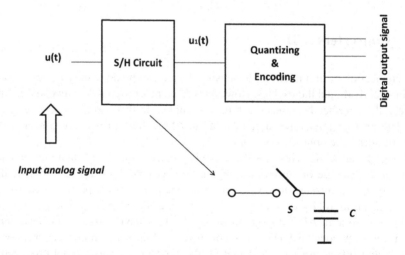

Fig. 2.10 Outline of analog to digital converter

aperture jitter, transient response, and over voltage recovery. Most converters sample with 6–24 bits of resolution, and produce fewer than 1 Mega samples per second. Thermal noise generated by passive components such as resistors masks the measurement when higher resolution is desired.

All ADCs suffer from nonlinearity errors caused by their physical imperfections, permitting their output to deviate from a linear function (or some other function, in the case of a deliberately nonlinear ADC) of their input. These errors can sometimes

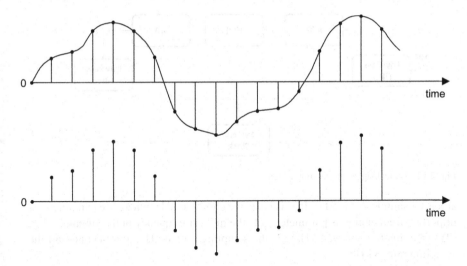

Fig. 2.11 Sampling Theorem

be mitigated by calibration, or prevented by testing. Important parameters are integral nonlinearity (INL) and differential nonlinearity (DNL). These reduce the dynamic range of the analog signals that can be digitized by the ADC, also reducing the effective resolution of the ADC.

The cost of an ADC is proportional to the following parameters: accuracy, number of bits, and stability. But even the most expensive ADC can compromise accuracy when excessive noise interferes with the input signal, whether that signal is in millivolts or much larger.

One technique for reducing noise and ensuring measurement accuracy is to eliminate ground loops that can occur when two or more devices are connected to ground terminals at different locations.

The input signal is continuous in time and is necessary to define a rate according to the extrapolation of new digital values from analog signal. The rate of new value is called sampling rate or sampling frequency of the converter (Sample and Hold, Fig. 2.11).

To process analog signals in computers, we need to convert the signals to "digital" form. To convert an analog signal to a digital form it must first be band-limited and then sampled. Theoretically the maximum frequency that can be useful is half the sampling frequency.

The sampled signals are represented by multiples of sampling period, T, as $s(nT)$ where n is an integer. Typically, an ADC is used to convert voltage (sampling value) to a digital number corresponding to a certain voltage level (Fig. 2.12). The process may be reversed through a DAC.

The sampling theorem indicates that an analog signal can be properly sampled, only if its maximum frequency is not above half of the sampling rate. If frequency components are above this limit, they will be aliased (Aliasing). We consider an analog signal composed of frequencies between DC to 2 kHz. It must be sampling

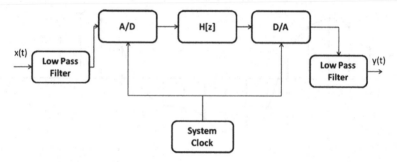

Fig. 2.12 Typical signal processing system

at least 4 samples/s (4 kHz). In this case, we choose 6 kHz, where we can note four important frequencies as parameters: (1) the highest frequency in the signal, 2 kHz; (2) twice this frequency, 4 kHz; (3) the sampling rate, 6 kHz; and (4) one-half the sampling rate, 3 kHz.

In mathematical terms, the sampling theorem (Nyquist theorem) indicates that if an analog signal $s(t)$ has a band limited, its fourier transform $Sa(i\omega)$ is given by

$$S_a(i\omega) = \int_{-\infty}^{+\infty} s(t) * \exp(-i\omega t)dt \tag{2.1}$$

such that

$$S_a(i\omega) = 0 \tag{2.2}$$

for

$$\omega \geq 2\pi W \tag{2.3}$$

Then the analog signal can be sampled with rate of period:

$$T \leq 1/2W \tag{2.4}$$

The quantity W is called the Nyquist frequency. If this rate is not verified, various types of distortion can be occurred; in particular:

- Aliasing. A precondition of the Nyquist theorem is that the signal must be bandlimited. However, in practice, no time-limited signal can be bandlimited. Designing a Sample-Hold with an appropriate guard band is possible to reduce aliasing effect.
- Jitter or deviation from the precise sample timing intervals [3, 6, 7].

The DS1843 of Maxim [11] is a sample-and-hold circuit (Figs. 2.13 and 2.14) useful for capturing fast signals where board space is constrained. It includes a differential, high-speed switched capacitor input sample stage, offset nulling circuitry, and an output buffer. The DS1843 is optimized for utilization in optical line transmission (OLT) systems for burst-mode RSSI measurement in conjunction with an external sense resistor. The input voltage is sampled using a 5 pF capacitor on the positive input and another on the negative input. The capacitors are connected to the input when SEN is high. In addition to the sampling capacitors, the inputs have

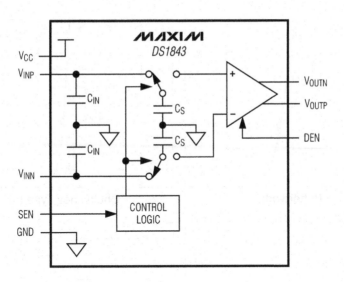

Fig. 2.13 Maxim DS1843 [8]

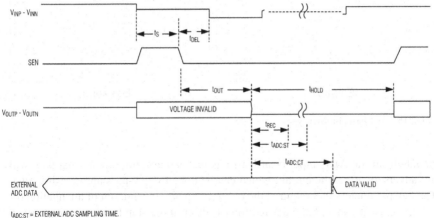

$t_{ADC:ST}$ = EXTERNAL ADC SAMPLING TIME.

$t_{ADC:CT}$ = EXTERNAL ADC CONVERSION TIME.

DEN IS CONNECTED TO V_{CC} FOR DIFFERENTIAL OUTPUT.

NOTE: THIS TIMING DIAGRAM IS APPLICABLE FOR SINGLE-ENDED AND DIFFERENTIAL OUTPUT CONFIGURATIONS.

Fig. 2.14 DS1843: Timing diagram [8]

parasitic capacitance (C_{IN}). These capacitors must be fully charged before SEN is switched to low in order to ensure the accurate sampling. An RC time constant is created by the resistance of the voltage source connected to the DS1843s input and the capacitances on this node.

An anti-aliasing filter is a filter used before a signal sampler, to restrict the bandwidth of a signal to approximately satisfy the sampling theorem. When selecting a filter, the goal is to provide a cutoff frequency that removes unwanted

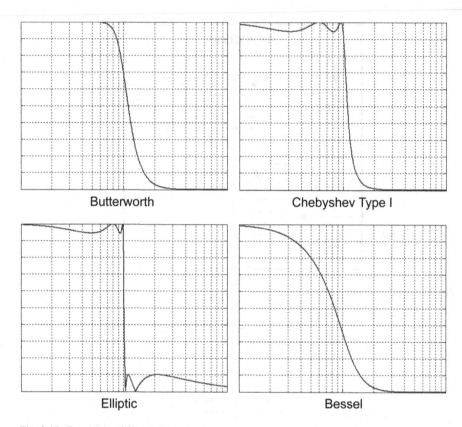

Fig. 2.15 Four basic of filter types

signals from the ADC input or at least attenuates them to the point that they will not adversely affect the circuit; which can be done by low-pass anti-aliasing filter. The key parameters that required the observation are: the amount of attenuation (or ripple) in the pass band, the desired filter roll off in the stop band, the steepness in the transition region, and the phase relationship of the different frequencies as they pass through the filter (Fig. 2.15) [5, 6, 8].

2.4.1 Parameters

ADC performance specifications are generally categorized in two ways: DC accuracy and dynamic performance. Most applications use ADCs to acquire signal like temperature, force, and so on. Each designer will consider the most important factor to project a specific and efficient ADC. Some important factors can be described in the following points:

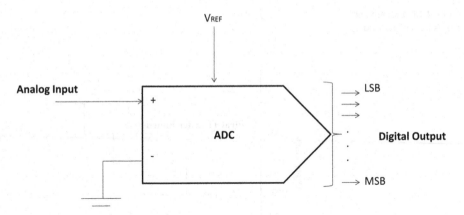

Fig. 2.16 Single ended conversion mode

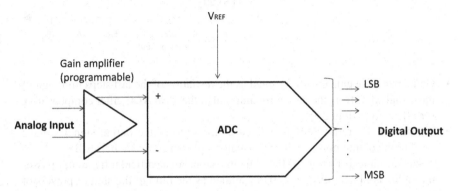

Fig. 2.17 Differential conversion mode

- Input voltage range: The input voltage range of an ADC is determined by the reference voltage (V_{REF}) applied to the ADC.
- Resolution: The resolution, expressed in bits, of the converter indicates the number of discrete values (binary) that can be produced over the range of analog values.
- Quantization error: Quantization error is an artifact of representing an analog signal with a digital number (in other words, an artifact of analog-to-digital conversion).
- Conversion mode: A conversion mode is a method in which the ADC processes the input. A standard ADC has two types of conversion modes: single ended conversion mode (Fig. 2.16) and differential conversion mode (Fig. 2.17).
- Offset error: The offset error is defined as the deviation of the actual ADCs transfer function from the perfect ADCs transfer function related at the point of LSB (Fig. 2.18).

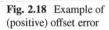

Fig. 2.18 Example of (positive) offset error

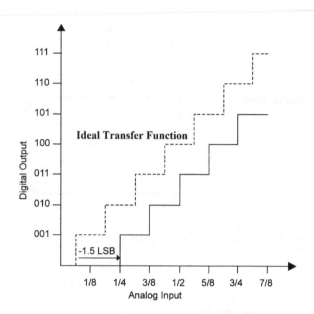

- Gain error: The gain error is defined as the deviation of the last steps midpoint of the actual ADC from the last steps midpoint of the ideal ADC, after compensating for offset error (Fig. 2.19).
- Signal-to-noise-ratio: SNR is defined as the ratio of the output signal voltage level to the output noise level. It is usually represented in decibels (dB)
- Total harmonic distortion (THD): An input signal of a particular frequency passes through a nonlinear device. THD is defined as the ratio of the sum of powers of the harmonic frequency components to the power of the fundamental/original frequency component.
- Impedances and capacitances of ADC: From Fig. 2.20 the input impedance of the ADC is the combination of R_{ADC} and the impedance of the capacitor C_{ADC} that is also called as sampling capacitor.

2.4.2 Successive-Approximation ADC

One method of ADC is called successive-approximation (Fig. 2.21). The successive approximation ADC circuit typically consists of four sub circuits:

1. An S-H circuit to sample the input voltage.
2. An analog voltage comparator that compares Vi with the output of the DAC and sends the result to the successive approximation register (SAR).
3. An SAR sub-circuit designed to supply an approximate digital code of Vin to the DAC.
4. An internal reference DAC supplies the comparator with an analog voltage equivalent of the digital code output of the SAR for comparison with Vi.

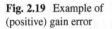

Fig. 2.19 Example of (positive) gain error

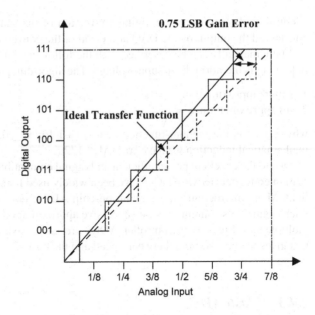

Fig. 2.20 Impedances and capacitances of ADC

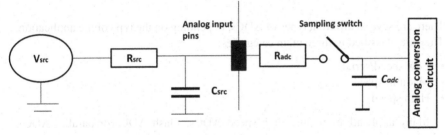

Fig. 2.21 Successive-approximation ADC

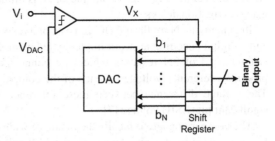

The SAR is initialized so that the most significant bit (MSB) is equal to a digital 1. Then, the code is sent into DAQ which supplies the analog equivalent of this code (Vref/2) into comparator with sampled input voltage. If this analog voltage is more bigger than Vin, the comparator causes the SAR to reset this bit. Then same test is done, repeating this process until every bit in the SAR is completed. The resulting codes represent the digital conversion of the sampled input voltage and considered the output by the DAQ at the end of the conversion (EOC).

Mathematically can be described with the following equation: Vi = x Vref, so x in $[-1, 1]$ is the normalized input voltage. The algorithm proceeds as follows:

1. Initial approximation $x_0 = 0$.
2. ith approximation $x_i = x_{i-1} - s(x_{i-1} - x)/2i$.

where, $s(x)$ is the signum-function $sgn(x)$ ($+1$ for x ≥ 0, -1 for x < 0). Using mathematical induction we have $|xn - x| \leq 1/2n$.

Due to the excellent power efficient and digital compatibility, successive approximation converters (SAR ADCs) have been widely used for some applications in this field. Many microcontrollers contain on-chip ADCs. For example the PIC16C7xx microcontrollers contain an 8-bit successive approximation ADC with analog input multiplexers. Most microcontroller ADCs are successive approximation because this gives the best solution between speed and the cost.

2.4.3 Flash ADC

There are several different types of ADCs, depending on the type of the application. They are classified in three main categories:

- Low speed/serial
- Medium speed
- High speed

Many applications require high speed ADCs, Flash ADC (or parallel ADC), which can be a good solution in this field.

The conversion speed in flash ADC is only one clock cycle, hence is the fastest ADC architecture (Fig. 2.22) available and it is limited only by both comparator and gate propagation delays.

In general, an N bit flash ADC consists of a resistors string, a set of comparators, and a digital encoding network implemented using XOR gates. The resistors string is composed of 2N resistors which are connected between Ref+ and Ref− to produce a reference voltage for each of the comparators as visualized in Fig. 2.22. The voltage difference between these reference voltages is equal to the least significant bit (LSB) voltage [7].

The comparators are typically a cascade of wide band low-gain stages. They are low gain because at high frequencies it is difficult to obtain both wide bandwidth and high gain. The comparators are designed for low-voltage offset, so that the input offset of each comparator is smaller than an LSB of the ADC. Otherwise, the comparator's offset could falsely trip the comparator, resulting in a digital output code that is not representative of a thermometer code. A regenerative latch at each comparator output stores the result. The latch has positive feedback, so that the end state is forced to either a 1 or a 0.

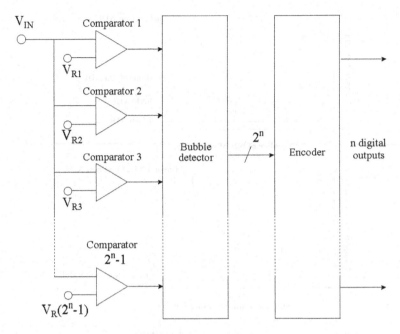

Fig. 2.22 Flash ADC

In an SAR converter, a single high-speed, high-accuracy comparator determines the bits, one bit at a time (from the MSB down to the LSB). This is done by comparing the analog input with a DAC whose output is updated by previously decided bits and thus successively approximates the analog input. This serial nature of the SAR limits its speed to no more than a few mega-samples per second (Msps), while flash ADCs exceed giga-samples per second (Gsps) conversion rates.

SAR converters are available in resolutions up to 16 bits. An example of such a device is the MAX1132. Flash ADCs are typically limited to around 8 bits. The slower speed also allows the SAR ADC to be much lower in power. The SAR architecture is also less expensive. Package sizes are larger for flash converters. In addition to a larger die size requiring a larger package, the package needs to dissipate considerable power and needs many pins for power and ground signal integrity (Fig. 2.23) [9].

Flash ADCs have been implemented in many technologies, from silicon-based bipolar (BJT) to complementary metal oxide FETs (CMOS) technologies. An additional advantage of the flash converter, often overlooked, is the ability for it to produce a nonlinear output.

A pipelined ADC works with a parallel structure where each stage works on one to a few bits of successive samples concurrently. This design improves speed of the ADC. This pipelined ADC requires accurate amplification stage in the DACs and interstages amplifiers [6, 9].

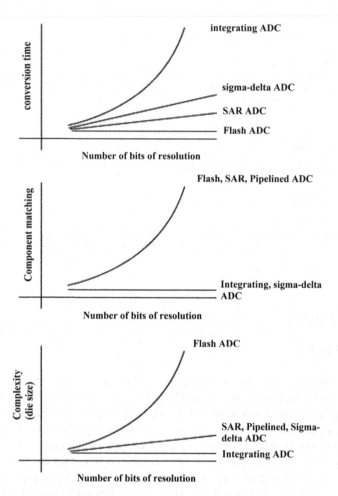

Fig. 2.23 Architectural trade-off

2.4.4 8-Bit, 500 Msps Flash ADC of Maxim

The MAX1150 [9] is a parallel flash ADC capable of digitizing full-scale (0 V to −2 V) inputs into 8-bit digital words at an update rate of 500 Msps (Figs. 2.24 and 2.25). The ECL-compatible outputs are demuxed into two separate output banks, each with differential data-ready outputs to ease the task of data capture. The MAX1150s (Fig. 2.26) wide input bandwidth and low capacitance eliminate the need for external track/hold amplifiers for most applications: DAQ systems, radar, digital oscilloscope. A major advance over previous flash converters is the inclusion of 255 input preamplifiers between the reference ladder and input comparators. The preamplifiers act as buffers to stabilize the input capacitance so that it remains constant over different input voltage and frequency ranges, making the part easier

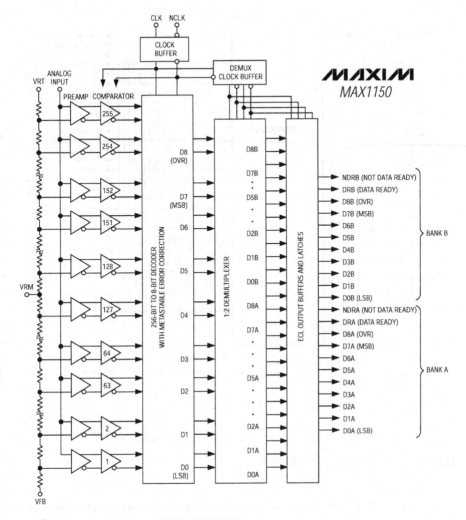

Fig. 2.24 Functional Diagram of MAX1150 [10]

to drive than previous flash converters. The circuit of Fig. 2.25 shows a method of achieving the least error by correcting for integral linearity, input-induced distortion, and power-supply/ground noise (Fig. 2.23) [10].

2.5 Converters D/A

A DAC is an electronic device that converts a digital (usually binary) signal to an analog signal (current, voltage, or electric charge). A common use of DACs is the generation of audio signals from digital information in music players.

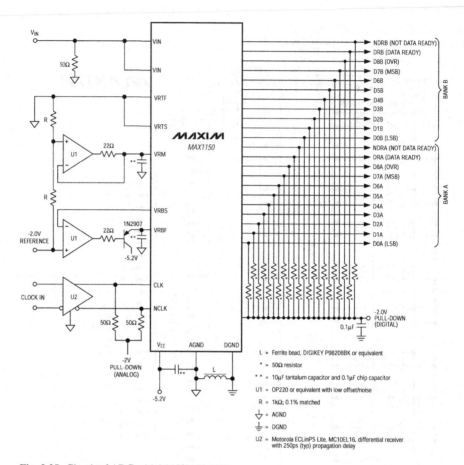

Fig. 2.25 Circuit of ADC with MAX1150 [10]

DACs (Fig. 2.28) are almost exclusively used on integrated circuits; there are many DAC architectures which have different advantages and disadvantages. The choice of a particular DAC for an application is determined by a variety of measurements including some parameters such as speed and resolution [6].

2.5.1 Parameters

DACs are very important to system performance. The most important characteristics of these devices are:

- Resolution: This is the number of possible output levels the DAC is designed to reproduce.

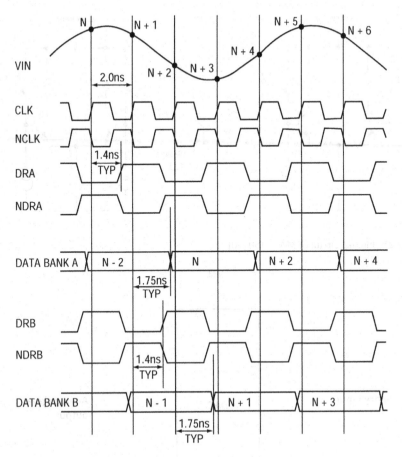

Fig. 2.26 Timing diagram of MAX1150 [10]

- Monotonicity: Ability of a DAC's analog output to move only in the direction that the digital input moves.
- Full scale range (FSR): Maximum output signal for the DAC.
- SFDR, Spurious Free Dynamic Range: The difference between the RMS power of the fundamental frequency and the largest spurious signal in the bandwidth.
- Glitch (Fig. 2.29): When a DAC is set to a desired output code, it does not immediately converted to the specific output current or voltage. It changes to the new output value over a finite time (the settling time) and that can cause signal spikes (glitches) rather than transitioning from old to new output values smoothly and monotonically. Because the glitch occurs right after a DAC is updated and disappeared within the first few microseconds, if the DAC output and output buffer amplifier input are decoupled when the DAC is updated and stay decoupled until the glitch disappears, the glitch will not pass through the output buffer amplifier. As shown in Fig. 2.30, this solution uses an S/H concept

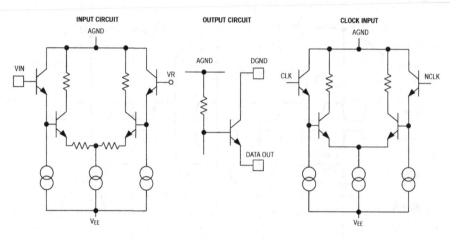

Fig. 2.27 Internal circuit of MAX1150 [10]

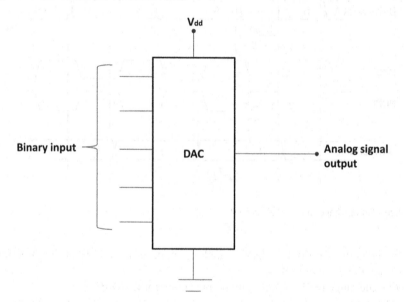

Fig. 2.28 Converter D/A

to eliminate glitches. Before the DAC is updated, switch SW1 is closed. The sampling capacitor samples the DC level of the previous DAC code. During a digital code transition, as the DAC is being updated, the switch is opened and the capacitor (CH) holds the DC level of the previous DAC code. The amplifier output is maintained at this DC level while the glitch occurs. After the glitch disappears, the switch closes again. The value of the T/H capacitor can be much smaller, because this capacitor is used to hold the DC level of the previous DAC code, as opposed to reducing the amplitude of a glitch. Small glitches

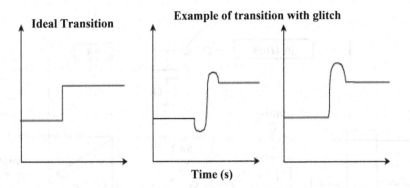

Fig. 2.29 Glitch

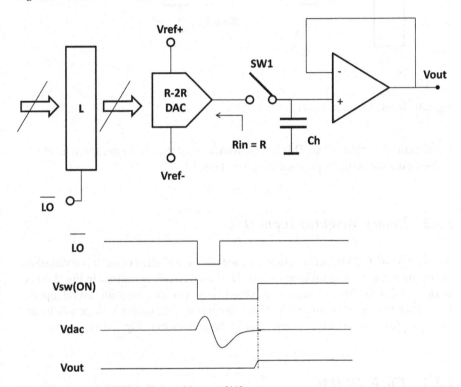

Fig. 2.30 Integrated T/H deglitch architecture [11]

can still occur when the T/H switch is turned on or off due to charge sharing and injection, but the associated glitch amplitude is much smaller. An improved Deglitches Circuit can be visualized in Fig. 2.31: To eliminate the base current of the sampling switch, differential charge cancellation is being used [11].

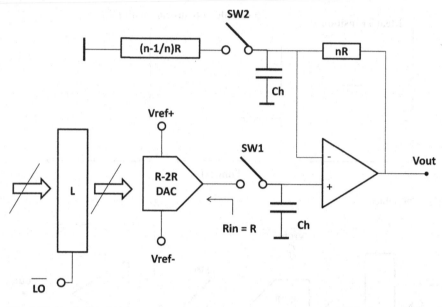

Fig. 2.31 Deglitch architecture [11]

- Maximum sampling rate: This is a measurement of the maximum speed at which the DACs can work to produce the correct output.

2.5.2   Binary-Weighted-Input DAC

An example of the DAC architecture is called binary-weighted input; it is a variation on the inverting summer op-amp circuit. The based circuit visualized in Fig. 2.32 is an operational amplifier in negative feedback to control the gain with several inputs. If we drive the inputs of this circuit with digital gates, the output voltage will be an analog representation of the binary value of these three bits (Fig. 2.33).

2.5.3   The R-2R DAC

An alternative to the binary-weighted-input DAC is the R/2R DAC (Fig. 2.34), which uses fewer unique resistor values. A disadvantage of the previous DAC design can be identified in different and precise input resistor values.

In an R-2R DAC design with supply voltages exceeding ±5 V, large voltage glitches (up to 1.5 V) can occur during the DAC's major-carry transitions. These glitches can propagate through the buffer amplifier and appear at output. The

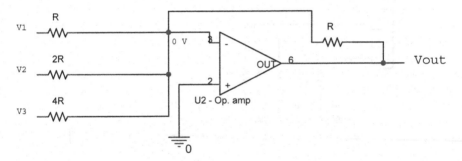

Fig. 2.32 Binary-weighted-input DAC

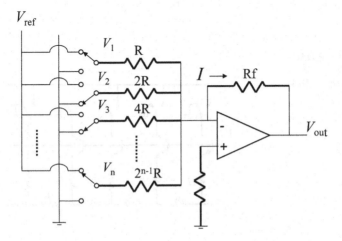

Fig. 2.33 Binary-weighted-input DAC

slewing of the level shifters that control the top (VREF+) and bottom (VREF−) single-pole double-throw switches causes the glitches (Fig. 2.35) [11].

2.5.4   8-Bit DACs with 2-Wire Serial Interface of Maxim

The MAX5109 [12] dual 8-bit DACs feature nonvolatile registers (Figs. 2.36 and 2.37). The MAX5109 has independent high and low reference inputs allowing maximum output voltage range flexibility. The reference rails accept voltage inputs that range from ground to the positive supply rail.

Applications:

- Digital gain and offset adjustments
- Programmable attenuators
- Portable instruments

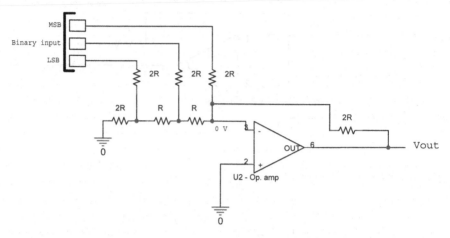

Fig. 2.34 R-2R DAC

Fig. 2.35 The R-2R DAC: simplified circuit

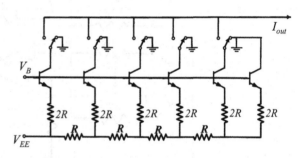

The MAX5109 uses a DAC matrix decoding architecture that saves power. A resistor string divides the difference between the external reference voltages, V_{REFH} and V_{REFL}. Row and column decoders select the appropriate tap from the resistor string, providing the equivalent analog voltage. The resistor string presents a code-independent input impedance to the reference and guarantees a monotonic output. Figure 2.38 shows a simplified diagram of one DAC [12].

The MAX5109 (Fig. 2.39) 8-bit DACs feature internal, nonvolatile registers that store the DAC states for initialization during power-up. This device consists of resistor-string DACs, rail-to-rail output buffers, a shift register, poweron reset (POR) circuitry, and volatile and nonvolatile memory registers (Fig. 2.38). The shift register decodes the control and address bits, routing the data to the proper registers. Writing data to a selected volatile register immediately updates the DAC outputs. The volatile registers retain data as long as the device is powered. Removing power clears the volatile registers. The nonvolatile registers retain data even after power is removed. On startup, when power is first applied, data from the nonvolatile registers is transferred to the volatile registers to automatically initialize the device. Read data from the nonvolatile or volatile registers using the 2-wire serial interface [12].

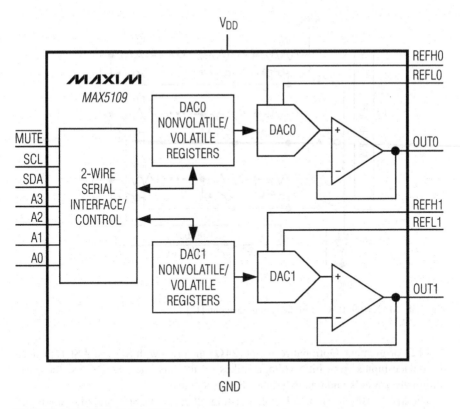

Fig. 2.36 MAX5109: Outline [12]

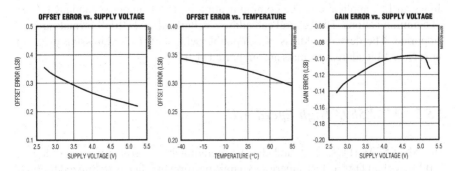

Fig. 2.37 MAX5109: Typical operating characteristics [12]

The MAX5109 features an I2C-compatible, 2-wire serial interface consisting of a bidirectional serial data line (SDA) and a serial clock line (SCL). SDA and SCL facilitate bidirectional communication between the MAX5109 and the master at rates up to 400 kHz (Fig. 2.40). The master (typically a microcontroller) initiates data transfer on the bus and generates SCL. SDA and SCL require pullup resistors

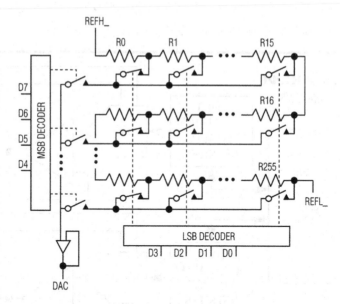

Fig. 2.38 MAX5109: DAC simplified circuit diagram [12]

(2.4 kΩ or greater). Optional resistors (24Ω) in series with SDA and SCL protect the device inputs from high-voltage spikes on the bus lines. Series resistors also minimize crosstalk and undershoot of the bus signals.

The MAX5109 is compatible with existing I2C systems. SCL and SDA are high-impedance inputs; SDA has an open-drain output. The typical operating circuit shows an I2C application. The communication protocol supports standard I2C 8-bit communications. The general call address is ignored, and CBUS formats are not supported. The devices address is compatible with 7-bit I2C addressing protocol only. No 10-bit address formats are supported [12].

2.5.5 MAX5893 High Speed A/D

The MAX5893 (Figs. 2.41 and 2.42) programmable interpolating, modulating, 500 Msps, dual DAC offers superior dynamic performance and is optimized for high performance wide band, single-carrier transmit applications. The device integrates a selectable 2x/4x/8x interpolating filter, a digital quadrature modulator, and dual 12-bit high-speed DACs on a single integrated circuit. At 30 MHz output frequency and 500 Msps update rate, the in-band SFDR is 84 dBc while consuming 1.1 W. The device also delivers 72 dB ACLR for single-carrier WCDMA at a 61.44 MHz output frequency. The selectable interpolating filters allow lower input data rates while taking advantage of the high DAC update rates. These linear-phase interpolation filters ease reconstruction filter requirements and enhance the pass band dynamic

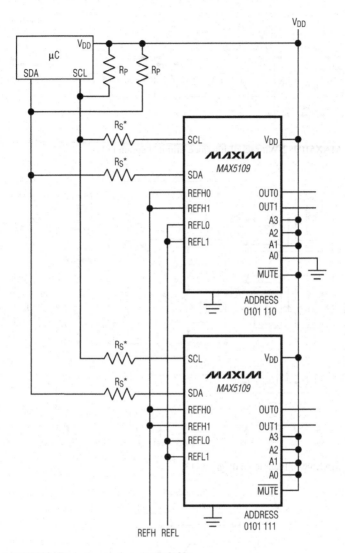

Fig. 2.39 MAX5109: Typical operating circuit [12]

performance. Individual offset and gain programmability allow the user to calibrate out local oscillator (LO) feedthrough and sideband suppression errors generated by analog quadrature modulators. The MAX5893 features a standard 1.8 V CMOS, 3.3 V tolerant data input bus for easy interface. A 3.3 V SPI port is provided for mode configuration. The programmable modes include the selection of 2x/4x/8x interpolating filters, $f_{IM}/2$, $f_{IM}/4$ or no digital quadrature modulation with image rejection, channel gain and offset adjustment, and offset binary or twos complement data interface [13].

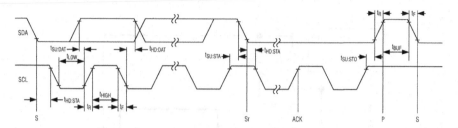

Fig. 2.40 MAX5109: 2-Wire serial-interface timing diagram [12]

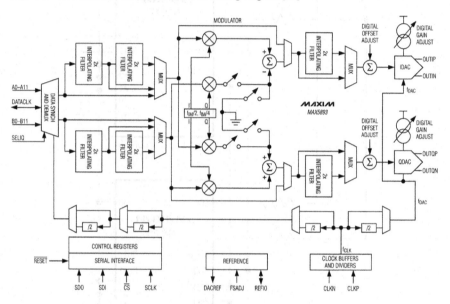

Fig. 2.41 MAX5893: Functional diagram [13]

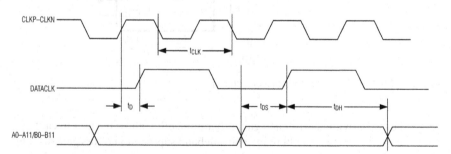

Fig. 2.42 MAX5893: Data input timing diagram [13]

2.6 Digital Signal Processing

Digital Signal Processing (DSP) is one of the core technologies in quickly growing. Applications can be found in wireless communications, audio and video processing, and industrial control. DSP has become a key electronic component in many of the medical and industrial products. In Math, DSP can be defined as manipulation of an information to modify or improve it in some way.

The general purpose of digital signal processors (Fig. 2.43) is dominated by applications in embedded. They also found application in cellular phones, modem, high definition television, and so on.

Some features of DSP are as follows:

• Efficient memory access
• Circular buffering: Circular buffers are used to store the most recent values of a continually updated signal
• Specialized instruction sets

DSP computes using sampled digital data; all applications require filter device. Filtering is done using finite impulse response (FIR) filter. The output of this filter is a weighted linear combination of current. The operation is described by the following equation, which defines the output sequence $y[n]$ in terms of its input sequence $x[n]$:

$$y[n] = \sum_{i=0}^{N} b_i x[n-i] \tag{2.5}$$

where:

• $x[n]$ is the input signal
• $y[n]$ is the output signal
• b_i are the filter coefficients, also known as tap weights, that make up the impulse response
• N is the filter order

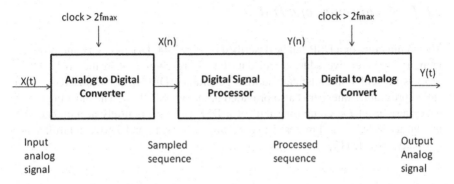

Fig. 2.43 Digital signal processing

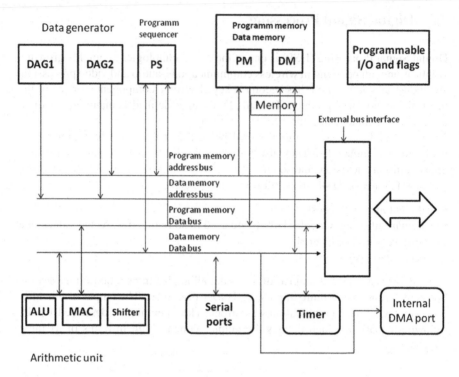

Fig. 2.44 Architecture of a DSP

The properties of an FIR can be described in the following points:

- Require no feedback
- Are inherently stable
- They can easily be designed to be linear phase

2.6.1 Architecture of a DSP

The architecture of a DSP can be described by Fig. 2.44. Data memory and program memory with relative address and data buses are connected to the ALU. DSPs have two calculation units: data address generators (DAGs). Besides DAGs, a unit called program sequencer is also provided in many DSP chips. The most important characteristic of the computing unit of a DSP is a fast multiply-accumulate unit known as MAC. On-chip serial and parallel I/O ports and DMA controller are normally provided [5].

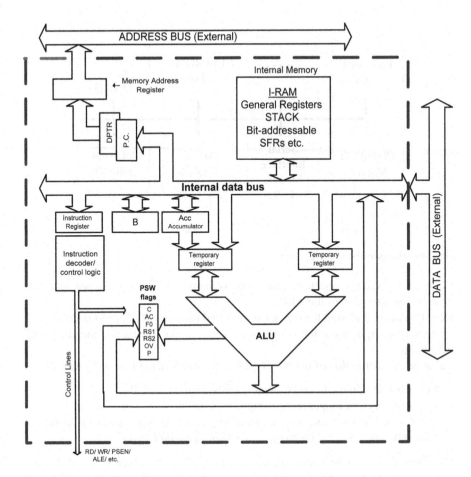

Fig. 2.45 Internal part of a microprocessor

2.7 Microprocessor and Microcontrollers

The functions of a computer's central processing unit (CPU) can be incorporated in a microprocessor. It is a programmable device that accepts digital data in input and can be defined as an example of sequential logic circuit (Fig. 2.45).

2.7.1 CPU Structure

The CPU can be composed of the following units:

Arithmetic and Logic Unit (ALU): The goal is the execution of operations such as addition, substraction, and so on.

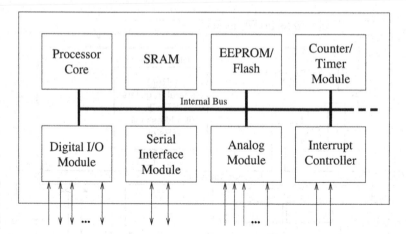

Fig. 2.46 Microcontroller: basic layout

Control Unit (CU): It checks the instructions in and out of the processor, and also controls the operation of the ALU.

Register Array: It is used for the fast storage and recovery of data and instructions.
System Bus: It is used for connections between the processor, memory, and peripherals.

Memory: An distinction of memory types can be made in the following ways:

* Register File: relatively small memory embedded on the CPU.
* Data Memory: for longer term storage.
* Instruction Memory: like the data memory, the instruction memory is usually a relatively large external memory (at least with general CPUs).

The instruction set is an important feature of any CPU. It influences the code size and much memory space the programm is contained. The metrics of the instructions set that are important for a design decision are: instruction size, execution speed, available instructions, and addressing modes.

2.7.2   Microcontrollers

A microcontroller (Fig. 2.46) differs from a microprocessor in many ways. The most important difference is its functionality. The microprocessor must be used with other components such as memory or components for data transfer. Simply, in order to communicate with peripheral environment, the microprocessor must use specialized devices. On the other hand, the microcontroller is designed to be all in one. As example, we consider the microcontroller of Maxim: MAXQ612/MAXQ622. In the Fig. 2.46 it is visualized the block diagram of a typical microcontroller. All components are connected via an internal bus and all devices are on chip:

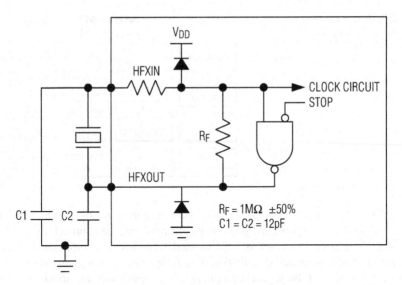

Fig. 2.47 Microcontroller MAXQ612/MAXQ622: on-chip oscillator [14]

- Processor Core: Arithmetic logic unit, the control unit, and the registers (stack pointer, program counter, accumulator register, register file, …)
- Memory
- Interrupt controller
- Digital I/O
- Interfaces
- Watchdog timer
- Debugging unit
- Timer/counter
- Interrupts

The basic purpose of any such interface is to allow microcontroller to communicate with other units, microcontrollers, peripherals, or a host PC:

- The Serial Communication Interface (SCI): asynchronous communication interface (Universal Asynchronous Receiver Transmitter, UART).
- The Serial Peripheral Interface (SPI) is a simple synchronous point-to-point interface based on a master–slave principle.
- The Inter-IC bus (IIC) is a synchronous bus that operates on a master–slave principle.

2.7.3 Microcontroller MAXQ612/622

The MAXQ612/MAXQ622 (Figs. 2.47, 2.48) is based on a low-power implementation of the new 16-bit MAXQ family of RISC cores. The core supports the Harvard memory architecture with separate internal 16-bit program and data address buses.

Fig. 2.48 Microcontroller
MAXQ612/MAXQ622:
block diagram [14]

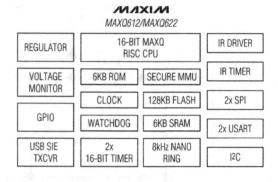

A fixed 16-bit instruction word is standard, but data can be arranged in 8 or 16 bits. The MAXQ core is a pipelined processor with performance approaching 1 MIPS per MHz. The 16-bit data path is implemented around register modules, and each register module contributes specific functions to the core. The accumulator module consists of sixteen 16-bit registers and is tightly coupled with the arithmetic logic unit (ALU). Program flow is supported by a configurable soft stack. Execution of instructions is triggered by data transfer between functional register modules or between a functional register module and memory. Because data movement involves only source and destination modules, circuit switching activities are limited to active modules only. For power-conscious applications, this approach localizes power dissipation and minimizes switching noise. The modular architecture also provides a maximum of flexibility and reusability that are important for a microprocessor used in embedded applications [14]. The microcontroller incorporates several memory types:

- 128 KB program flash memory
- 6 KB SRAM data memory
- 6 KB utility ROM
- Soft stack

A 16-bit-wide internal stack provides storage for program return addresses and can also be used for general purpose data storage. The internal watchdog timer greatly increases system reliability. The timer resets the device if software execution is disturbed. The watchdog timer is a free-running counter designed to be periodically reset by the application software. The dual-integrated SPI interfaces provide independent serial communication channels that communicate synchronously with peripheral devices in a multiple master or multiple slave system. The interface allows access to a 4-wire, full-duplex serial bus, and can be operated in either master mode or slave mode. Collision detection is provided when two or more masters attempt a data transfer at the same time. The microcontroller integrates an internal I2C bus master/ slave for communication with a wide variety of other I2C-enabled peripherals. The I2C bus is a 2-wire, bidirectional bus using two bus lines—the serial data line (SDA) and the SCL—and a ground line. Both the SDA and SDL lines must be driven as open collector/ drain outputs. External resistors are required to pull the lines to a logic-high state [14].

2.8 Amplifier

Signal amplifiers are electronic devices with the ability to amplify a relative small signal of a sensor (temperature sensors, magnetic field sensor, and so on). The quality of an amplifier can be described by a number of parameters, listed below:

- Gain: The ratio between output and input power or amplitude; it is usually measured in decibels.
- Bandwidth: The range of frequencies for which the amplifier works correctly.
- Noise: The noise level introduced in the amplification process.
- Slew rate: The maximum rate of voltage change per unit time.
- Overshoot: The output exceeds in its final value.

A particular amplifier is the feedback amplifier (Fig. 2.49): an amplifier which combines the output with the input so that a negative feedback opposes the original signal. Feedback in amplifiers gives better performance, in particular [6]:

- Increases the stability in the amplification
- Reduces distortion
- Increases the bandwidth of the amplifier

2.8.1 Design of Low-Noise Pre-amplifier

Low-noise amplifier (LNA) is an electronic amplifier used to amplify possibly very weak signals. A preamplifier is an electronic device which amplifies an analog signal. Generally is the stage that anticipates the high power amplifier.

In this example we design a simple preamplifier with background noise of the order of 0.8 nV/Hz@1 kHz. The circuit diagram is shown in Fig. 2.50. The main components used are LT1128 dell Linear Tech. and the JFET IF3602 of

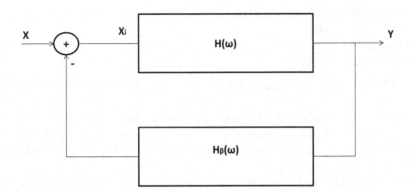

Fig. 2.49 Model of a feedback amplifier

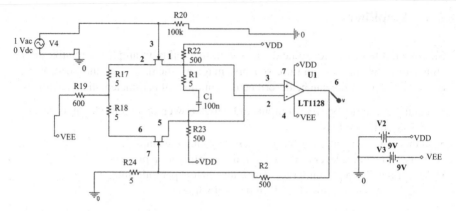

Fig. 2.50 Low noise preamplifier

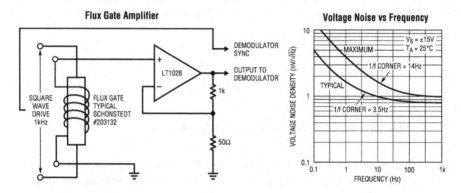

Fig. 2.51 Voltage noise LT1128 [15]

InterfetCorporation [15]. LT1128 (Figs. 2.51 and 2.52) is an operational amplifier ultra noise at high speed. Main characteristics are the following:

- Noise voltage: 0.85 nV/z@1 kHz
- Bandwidth: 13 MHz
- Slew rate: 5 V/uS
- Offset voltage: 40 uV

The IF3602, instead, is a Dual-N JFET used as stage for input of the operational amplifier.

The op-amp is one type of differential amplifier (Fig. 2.53). The inputs of the differential amplifiers consist of a V+ and a V− input, and ideally the op-amp amplifies only the difference in voltage between the two, which is called the differential input voltage [6].

The operational amplifier can be realized with bipolar junction transistor (BJT, as in the case of the LT1128) or MOSFET, which works at higher frequencies, with an

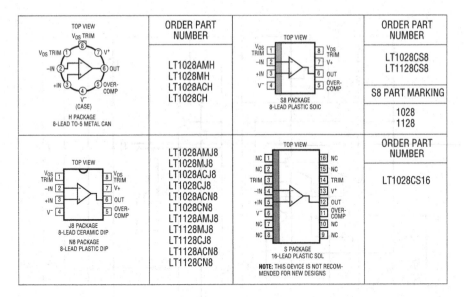

Fig. 2.52 Pin-outs of LT1128 [15]

input impedance higher and a lower energy consumption. The differential structure is used in those applications where it is necessary to eliminate the undesired common components to the two inputs. In this way, in output are eliminated eventual DC components on the input signal such as, for example, the thermal drift. The block diagram of the differential amplifier is shown in Fig. 2.54. We can define a differential gain ($A_d = A_2 - A_1$) and a common-mode gain ($A_c = A_1 + A_2/2$). An important parameter is the CMRR which is the ratio of common-mode gain to the differential-mode gain. This parameter is used to measure the performance of the differential amplifier. The differential circuit proposed is shown in Fig. 2.55. In this case the gain of mode-common is about $R_D/2R_1 + R_s + r_s$, while the differential gain is about $R_d/R_s + r_s$.

The circuit proposed in Fig. 2.50 is analyzed with P-spice simulator. We consider separately two principal blocks:

- The network $R_1 - C_1$
- The network of the feedback resistor

Feedback resistor has been regulated so as to obtain a gain of about 100. The maximum gain that the circuit can give is about 200. It is not recommended to use a potentiometer for adjusting the gain because it introduces noise to the system (Fig. 2.56).

The network $R_1 - C_1$ has the feature to adjust the band of the system; by varying R1 and having constant at 10 nF C1, we have obtained a series of curves that represented the response curve of the system in frequency (Fig. 2.57). Similar tests with R_1 constant at 10 ohms, and C_1 variable (Fig. 2.58).

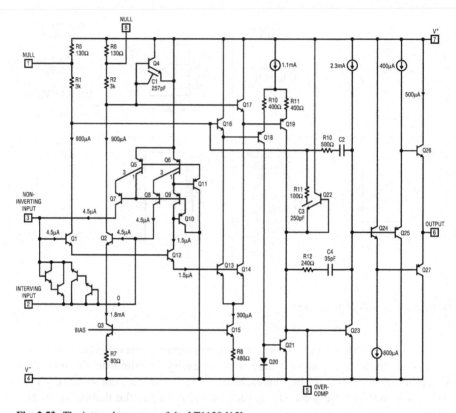

Fig. 2.53 The internal structure of the LT1128 [15]

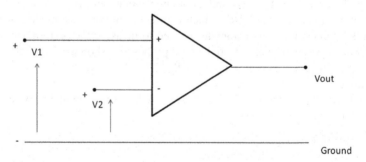

Fig. 2.54 Differential amplifier

The flexibility of the system can facilitate the choice of the response in frequency (Figs. 2.59, 2.60 and 2.61). The response curve of the system as function of the noise is shown in Fig. 2.62. The noises of internal current and voltage of the amplifiers depend on the intrinsic physical phenomena and are by their nature random, aperiodic, and uncorrelated. Typically have a distribution of amplitude of

Fig. 2.55 Differential circuit (JFET)

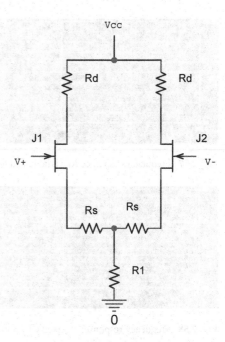

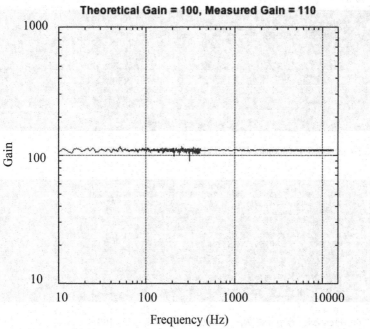

Fig. 2.56 Transfer function

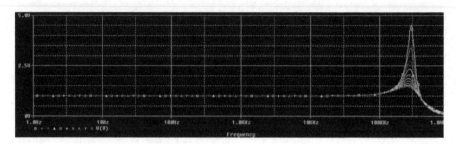

Fig. 2.57 Frequency response: R_1 variable and C_1 fixed

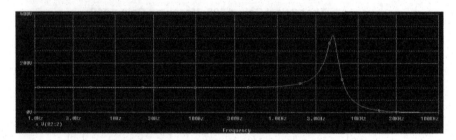

Fig. 2.58 Frequency response: R_1 fixed e C_1 variable

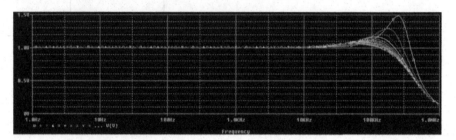

Fig. 2.59 Frequency response: $R_1 = 15 \ \Omega, C_1 = 100$ nF $(G = 100)$

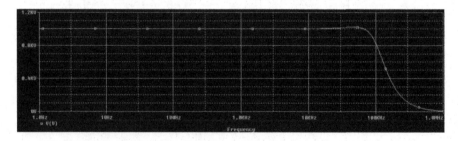

Fig. 2.60 Frequency response: $R_1 = 100 \ \Omega, C_1 = 10$ nF $(G = 100)$

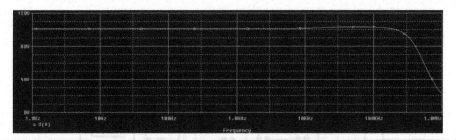

Fig. 2.61 Frequency response: $R_1 = 1\,k\Omega$, $C_1 = 10\,nF$ ($G = 100$)

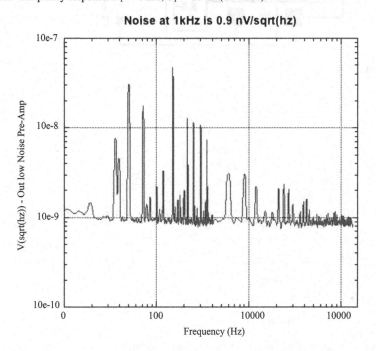

Fig. 2.62 System noise

Gaussian. The relationship between the peak to peak value and the effective value of these components is statistical. A possible qualitative rule is that the RMS value multiplied by 6 does not exceed the peak to peak in the 99.73 % of cases. In the evaluation of noise (0.8 nV/Hz to the frequency of about 1 kHz) in the operational is used to report all sources of noise at the input. In Fig. 2.63 the PCB of the circuit (Fig. 2.64) is visualized, and it uses SMD resistive and capacitive components to obtain a mini preamplifier.

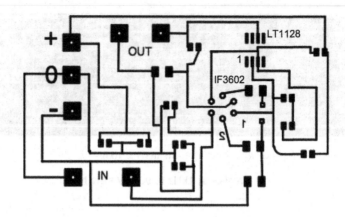

Fig. 2.63 PCB

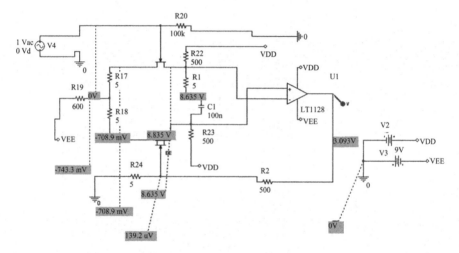

Fig. 2.64 Simulation of the circuit: values of voltages and currents

2.8.2 Low Noise Amplifier of the Maxim Integrated, MAX9632

The MAX9632 (Figs. 2.65, 2.66, 2.67, and 2.68) is a low-noise, precision, wide-band operational amplifier that can operate in a very wide +4.5 V to +36 V supply voltage range. The IC operates in dual (±18 V) mode. The IC is designed for extremely low-noise applications such as professional audio equipment, very high performance instrumentations, automated test equipment, and medical imaging. The low noise, combined with fast settling time, makes it ideal to drive high-resolution sigma delta or SARs ADCs. The IC is also designed for ultra-low-distortion performance [16].

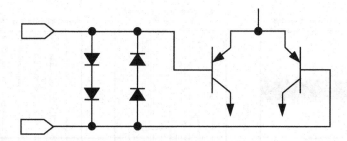

Fig. 2.65 MAX9632: input protection circuit [16]

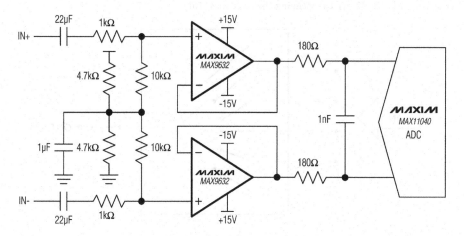

Fig. 2.66 MAX9632: typical application circuit [16]

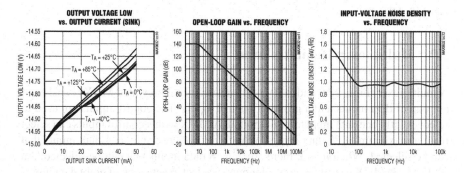

Fig. 2.67 MAX9632: typical operating characteristics [16]

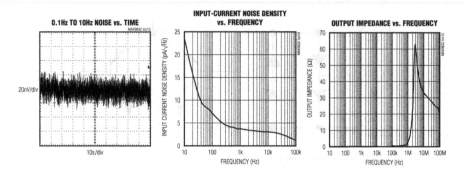

Fig. 2.68 MAX9632: typical operating characteristics [16]

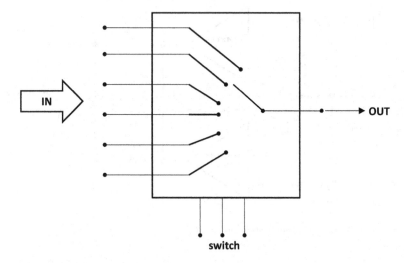

Fig. 2.69 MUX 8x1

2.9 Multiplexer/Demultiplexer

A multiplexer (Figs. 2.69 and 2.70) or data selector (abbreviated MUX) consists of a group of data inputs and a group of control inputs. The control inputs are used to select exactly one data input to be outputted. A MUX with n control inputs can select from a maximum of 2^n data inputs. When $n = 2$, there are $2^2 = 4$ data inputs that can be selected. When $n = 3$, there are $2^3 = 8$ data inputs that can be selected. An 8X1 ($n = 3$) MUX uses three control inputs to select exactly one of eight data inputs to be outputted. The three control inputs are labeled as A, B, and C (Fig. 2.69). On the other hand, a demultiplexer (or demux) is a device that sends in one of many data-output-lines a single input signal and getting out it in one of many data-output-lines [6].

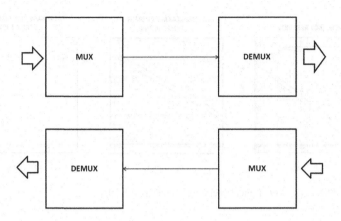

Fig. 2.70 MUX–DEMUX

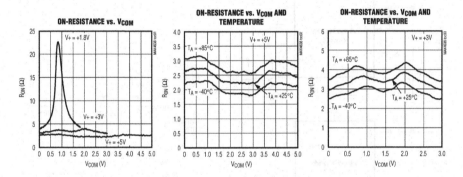

Fig. 2.71 MAX4638/4639: Typical operating characteristics [17]

2.9.1 Multiplexer/Demultiplexer of Maxim Integrated, MAX4638/4639

The MAX4638/MAX4639 (Figs. 2.71, 2.72, 2.73, 2.74, and 2.75) are low-voltage, CMOS analog muxes. The MAX4638 is an 8:1 mux that switches one of eight inputs (NO1–NO8) to a common output (COM) as determined by the 3-bit binary inputs A0, A1, and A2. The MAX4639 is a 4:1 dual mux that switches one of four differential inputs to a common differential output as determined by the 2-bit binary inputs A0 and A1. Both the MAX4638/MAX4639 have an EN input that can be used to enable or disable the device. When disabled all channels are switched off.

A +1.8 V to +5.5 V operating range makes the MAX4638/MAX4639 ideal for battery-powered, portable instruments. All channels guarantee break-before-make switching. All control inputs are TTL/CMOS-logic compatible. Decoding is in standard BCD format, and an enable input is provided to simplify cascading

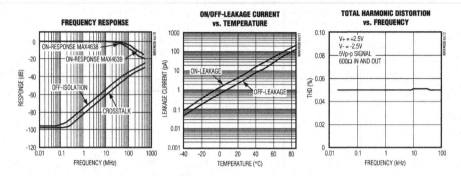

Fig. 2.72 MAX4638/4639: Typical operating characteristics [17]

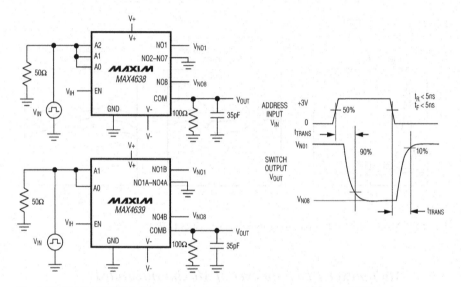

Fig. 2.73 MAX4638/4639: Test circuit and timing diagram (transition time) [17]

of devices. These devices are available in small 16-pin TQFN, TSSOP, and SO packages, as well as a 20-pin TQFN package [17].

Proper power-supply sequencing is recommended for all CMOS devices. Always sequence V+ on first, then V−, followed by the logic inputs. If power-supply sequencing is not possible, add two small-signal diodes (D1, D2) in series with the supply pins for over voltage protection (Fig. 2.76). Adding diodes reduces the analog signal range to one diode drop below V+ and one diode drop above V−, but does not affect the devices low switch resistance [17].

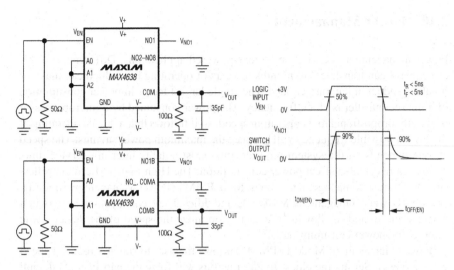

Fig. 2.74 MAX4638/4639: Test circuit and timing diagram (enabling time) [17]

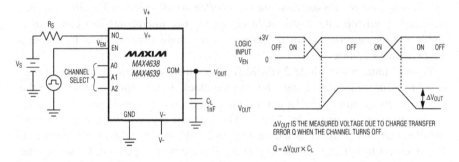

Fig. 2.75 MAX4638/4639: Test circuit and timing diagram (charge injection) [17]

Fig. 2.76 MAX4638/4639:
Over voltage protection using
external blocking diodes [17]

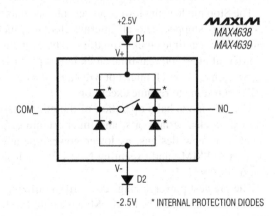

2.10   Power Management

Power management is critical in battery-powered applications. Differences of microamperes can translate into months or years of operating life, which can make or break a product in the marketplace. The single largest factor in power consumption of a microcontroller is clock frequency. The power consumed by a microprocessor is directly proportional to its operating speed, so it follows that a device operating at the lowest possible frequency will produce the maximum power savings. The speed chosen depends on the system requirements, most notably interrupt service time. Temperature can also affect power consumption. The High-Speed Microcontrollers support four clock management modes: Stop, PMM1 (Power Management Mode 1), PMM2 (Power Management Mode 2), and Idle. They can dynamically switch between these modes, allowing the user to optimize the speed of the device while minimizing power consumption.

Power Management Mode 1 (PMM1) allows the user to run at a reduced speed to save power. Setting the clock divider rate bits will force the part from its default 4 clocks per machine cycle (divide by 4) to 64 clocks per machine cycle (divide by 64). The external crystal continues to operate at full speed. All peripherals and instructions will operate at this reduced speed. The microcontroller can resume divide by 4 operation by setting the appropriate clock divider rate bits or by utilizing the switchback feature.

Power Management Mode 2 (PMM2) allows the user to run at an even slower speed to improve power savings. Setting the clock divider rate bits will force the part from its default 4 clocks per machine cycle (divide by 4) to 1,024 clocks per machine cycle (divide by 1,024). The external crystal continues to operate at full speed. All peripherals and instructions will operate at this reduced speed. The microcontroller can resume full-speed (divide by 4) operation by setting the appropriate clock divider rate bits or by utilizing the switchback feature. This mode permits an even greater power savings over PMM1.

The Stop mode is the lowest power state available. While in this mode the crystal oscillator is stopped, and all internal clocking, including the Watchdog Timer, is halted. The real time clock is unaffected by Stop mode. The Stop mode is exited by an external interrupt, real-time clock interrupt, an external reset via the RST pin, or a power-on reset. Each interrupt will cause the device to vector to the corresponding interrupt routine to resume execution.

The Idle mode halts operation of the microcontroller processor core but leaves internal clocks, serial ports, and timers running. Use of this mode is not recommended on new designs, as lower power operation can be achieved by placing the part in PMM2 and executing NOPs. Its inclusion provides backward software compatibility [19].

The greatest power savings come from utilizing the power management modes. Unlike other techniques, Power Management Modes 1 and 2 (PMM1 and PMM2) allow the user to reduce power consumption without sacrificing performance. Although the power management features are an important part of a power efficient

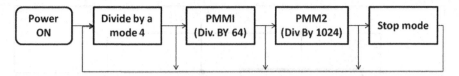

Fig. 2.77 Progression of clock speed modes

design, a thorough understanding of the microprocessor will allow the system designer to achieve maximum power savings. The clock speed management modes are designed to be part of a progressive level of power reduction, based on external activity and performance needs. PMM1 and PMM2 provide the lowest level of power consumption while still permitting full computational and peripheral operation. Figure 2.77 demonstrates the progression of clock management modes. As explained later, transitions between PMM1 and PMM2 must be made through divide by 4 mode [18, 19].

Numerous diverse and conflicting constraints burden the designer of small handheld products. Aside from the customary restrictions on size and weight, these constraints include cost limitations, strict time schedules, battery-life goals measured in weeks instead of hours, and host computers that are (sometimes) overtaxed with the demands of power management [18, 19].

Because power requirements for handheld applications vary widely with product use, no single "best" power source exists for these applications. A device used intermittently is more concerned with no-load quiescent current than with full-load efficiency, and can operate satisfactorily with alkaline batteries. Cell phones, however, must contend with high peak loads and frequent use. This mode of operation emphasizes conversion efficiency over quiescent current, so cell phones are better served with a rechargeable battery.

In handheld product design, size limitations often dictate the number of battery cells early in the process. This is frustrating to the electrical engineer, and a substantial constraint, since the number (and type) of cells allowed determines the operating-voltage range. This, in turn, strongly affects the cost and complexity of the power supply. High cell counts enable the use of linear regulators and simple circuitry at the cost of extra weight and limited efficiency. Low cell counts compel the use of a more costly switching regulator, but the low cost of the battery may justify this expense [18]. A design with four single-cell batteries often provides an attractive compromise between weight and operating life. That number is particularly popular for alkaline batteries because they are commonly sold in multiples of four. Four-cell power for 5 V circuitry presents a design challenge, however. As a battery discharges, the regulator must first step down and then step up. This requirement precludes use of the simpler, one-function regulator topologies that can only step down, step up, or invert. One effective solution to this problem is the single-ended primary inductance converter (SEPIC), in which VOUT is capacitively coupled to the switching circuitry (Fig. 2.78). The absence of

62   2 Data Acquisition Systems: Hardware

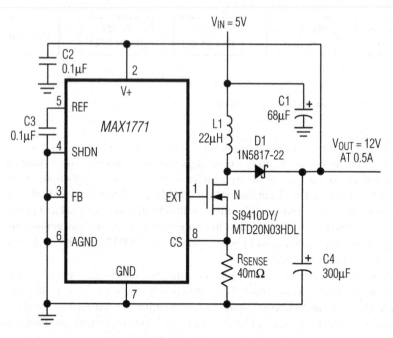

Fig. 2.78 MAX1771 step-up controller [18]

a transformer is one of several advantages that this configuration has over flyback-transformer regulators and combination step-up/linear regulators [18, 19].

2.10.1 Automotive Power-Management MAX16920

The MAX16920 (Figs. 2.79 and 2.80) power-management IC integrates three high-voltage step-down DC–DC converters, one high-voltage linear regulator, and an overvoltage protection block. The MAX16920 is optimized for high efficiency and low standby current. The power dissipation of the MAX16920 is made up of three components: power dissipation due to the DC–DC converters, power dissipation due to the linear regulator, and internal power dissipation [20].

2.10.2 Power-Management ICs for Single-Cell, MAX8662/MAX8663

The MAX8662/MAX8663 highly integrated PMICs are designed for use in smart cellular phones, PDAs, Internet appliances, and other portable devices (Fig. 2.81).

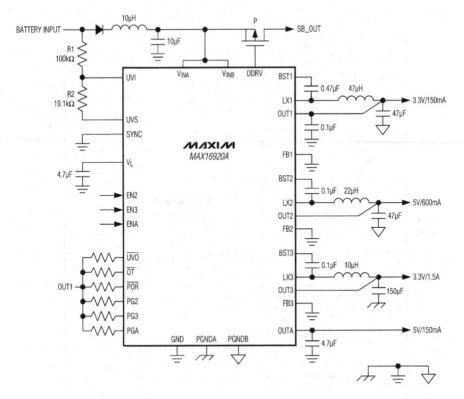

Fig. 2.79 MAX16920: Typical operating circuit [20]

They integrate two synchronous buck regulators, a boost regulator driving two to seven white LEDs (MAX8662 only), four low dropout (LDO) linear regulators, and a linear charger for a single-cell Li+ battery. Figure 2.81 is the block diagram and application circuit. SPS circuitry offers flexible power distribution between an AC adapter or USB source, battery, and system load, and makes the best use of available power from the AC adapter/USB input. The battery is charged with any available power not used by the system load. If a system load peak exceeds the current limit, supplemental current is taken from the battery. Thermal limiting prevents overheating by reducing power drawn from the input source. Two step-down DC–DC converters achieve excellent light-load efficiency and have on-chip soft-start circuitry; 1 MHz switching frequency allows for small external components. Four LDO linear regulators feature low quiescent current and operate from inputs as low as 1.7 V. This allows the LDOs to operate from the stepdown output voltage to improve efficiency. The white LED driver features easy adjustment of LED brightness and open-LED overvoltage protection. A 1-cell Li+ charger has programmable charge current up to 1.25A and a charge timer [21].

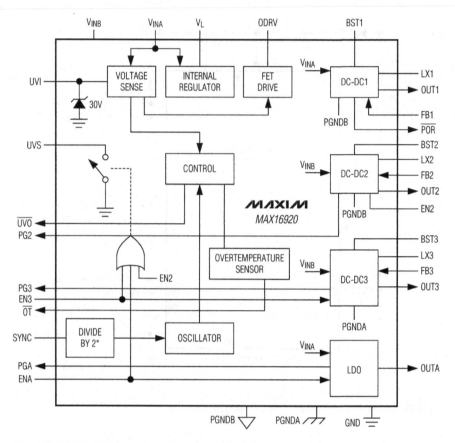

Fig. 2.80 MAX16920: Functional diagram [20]

2.11   Timing System

In the DAQ systems are required the design of timing and synchronization. The timing system management tasks between various electronics stages that required hardware synchronization. Example of timing system can be the following: handshaking, phase-lock looping, synchronizing an RF device, and so on. They are based on two main assumptions: all signals are digital and all components share a common and discrete notion of time; it is defined as clock signal located throughout the circuit.

The advantages of asynchronous circuits can be described below:

- Low power consumption
- High operating speed
- Better modularity

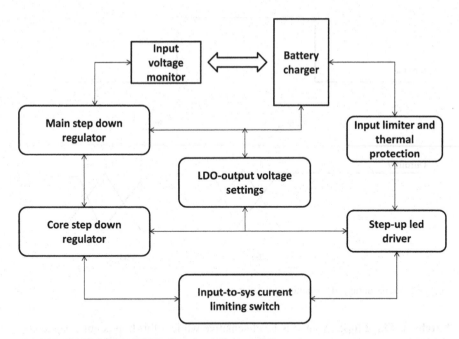

Fig. 2.81 MAX8662/MAX8663: Functional diagram [21]

- Robustness in terms of variations in supply voltage, temperature, and fabrication process parameters.

2.11.1 Timing Parameters for Combinational Logic

Timing characteristics can be defined by Fig. 2.82:

- Propagation delay (t_{pd}): The amount of time needed for a change in a logic input to result in a permanent change at an output.
- Contamination delay (t_{cd}): The amount of time needed for a change in a logic input to result in an initial change at an output [6].

2.11.2 Timing Parameters for Sequential Logic

Timing system with sequential circuits implemented certain timing characteristics that are specified in relation of the clock input.

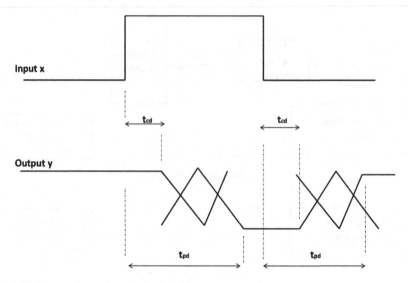

Fig. 2.82 Propagation and contamination delay

Latch vs. Flip-Flop: A latch is level-sensitive while a flip-flop is edge triggered. A latch stores digital data when the clock level is low and doesn't work when the level is high. A flip-flop, instead, stores digital data when the clock rises. Timing characteristics can be specified in relation to the rising (for positive edge-triggered) or falling (for negative-edge triggered) clock edge. In the following text we can specify parameters to explain sequential circuit behavior [7]:

- Propagation delay (t_{clk-q}): Time needed for a change in the flip-flop clock input D to result in a change at the flip-flop output Q.
- Contamination delay (t_{cd}): Time needed for a change in the flip-flop clock input to result in the initial change at the flip-flop output Q.
- Setup time (t_{su}): time before the clock edge that data input D must be stable the rising clock edge arrives.
- Hold time (t_{hold}): time after the clock edge arrives that data input D must be held stable in order for the flip-flop to latch the correct value.

2.11.3   Clock Skew and Clock Jitter

Clock skew is the spatial variation in arrival time of a clock transition (δ). It is a constant from cycle to another and given by $t_j - t_k$, where $t_j - t_k$ is the rising edge of the clock with respect to the reference between two points j and k (Figs. 2.83 and 2.84).

Clock jitter is related to the temporal variation of the clock period; the clock period can increase or reduce on a cycle-by-cycle basis (Fig. 2.85) [22].

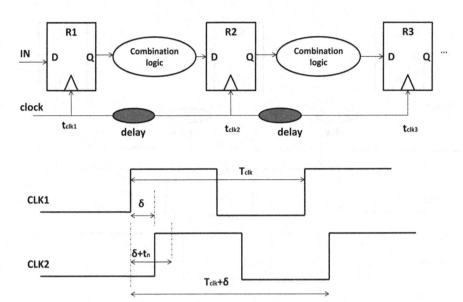

Fig. 2.83 Positive skew

2.11.4 MAX9155-Clock

The MAX9155 (Figs. 2.86 and 2.87) is a low-voltage differential signaling (LVDS) repeater, which accepts a single LVDS input and duplicates the signal at a single LVDS output. Its low-jitter, low-noise performance makes it ideal for buffering LVDS signals sent over long distances or noisy environments, such as cables and backplanes.

The MAX9155's tiny size makes it especially suitable for minimizing stub lengths in multidrop backplane applications. The SC70 package (half the size of a SOT23) allows the MAX9155 to be placed close to the connector, thereby minimizing stub lengths and reflections on the bus. The point-to-point connection between the MAX9155 output and the destination IC, such as an FPGA or ASIC, allows the destination IC to be located at greater distances from the bus connector [23].

2.12 Filtering

The most common filter are the Butterworth, Chebyshev, and Bessel (Figs. 2.88 and 2.89) types. Many other types are available, but 90 % of all applications can be solved with one of these three. Butterworth ensures a flat response in the pass-band and an adequate rate of roll-off; it is simple to understand and suitable for

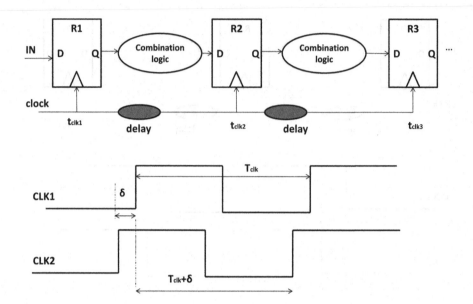

Fig. 2.84 Negative skew

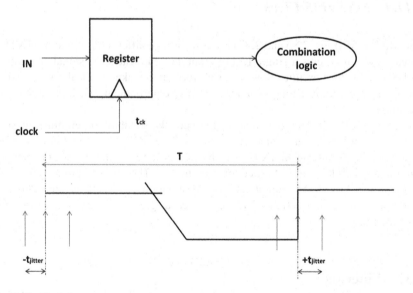

Fig. 2.85 Clock jitter

Fig. 2.86 Transition time and propagation delay test circuit [23]

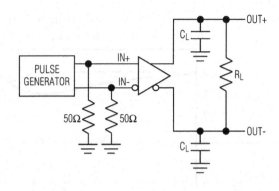

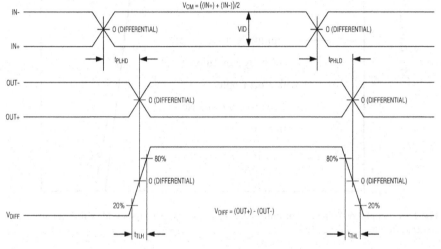

Fig. 2.87 Transition time and propagation delay timing diagram [23]

applications such as audio processing. The Chebyshev gives a much steeper roll-off, but pass-band ripple makes it unsuitable for audio systems. It is superior for applications in which the pass-band includes only one frequency of interest (e.g., the derivation of a sine wave from a square wave, by filtering out the harmonics). The Bessel filter gives a constant propagation delay across the input frequency spectrum. Therefore, applying a square wave (consisting of a fundamental and many harmonics) to the input of a Bessel filter yields an output square wave with no overshoot (all the frequencies are delayed by the same amount). Other filters delay the harmonics by different amounts, resulting in an overshoot on the output waveform. One other popular filter, the elliptical type, is a much more complicated filter that will not be discussed in this text. Similar to the Chebyshev response, it has ripple in the pass-band and severe roll-off at the expense of ripple in the stop-band [24].

In the Fig. 2.90 some examples of low-pass, high-pass, and generic filter design. Digital filters are algorithms used for digital computers. It is a linear operation

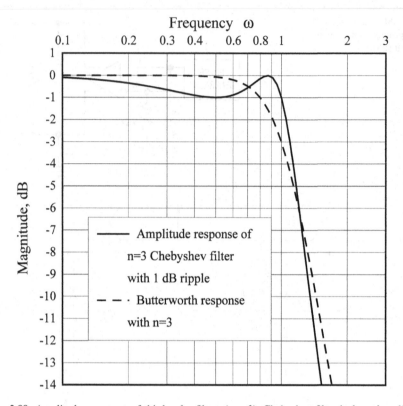

Fig. 2.88 Amplitude responses of third order filters ($n = 3$). Chebyshev filter in *boundary line*, Butterworth filter in *dashed line*

completed on sampled data. The main operations that includes are the following: smoothing (averaging), integrating, separating signals (filtering), and predicting.

Digital filters can be compared to analog filters. For example, low-pass have their digital equivalent but they have additionally some important features which make them well suited for digital communication system. Digital filters can be programmed and there are no impedance-matching problems and can have memories ifrequired

2.12.1 Digitally Programmed, Dual Second-Order Continuous Low-Pass Filter, MAX270/271

The MAX270/MAX271 (Figs. 2.91 and 2.92) are digitally programmed, dual second-order continuous-time low-pass filters. Their typical dynamic range of 96 dB surpasses most switched capacitor filters which require additional filtering to remove clock noise. The MAX270/MAX271 are ideal for anti-aliasing and DAC

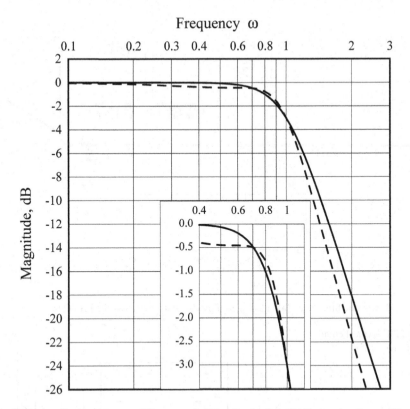

Fig. 2.89 Amplitude responses of Butterworth filter of order 3 (*solid line*) and optimum filter of order 3 (*dashed line*)

smoothing applications and can be cascaded for higher-order responses. Cutoff frequencies in the 1–25 kHz range can be selected [25].

Cut-Off Frequency: Cut-off frequency f_c is the frequency of 3 dB attenuation in the filter response. According to the data value in pin D0–D6, cut-off frequency is programmed from 1 to 25 kHz. The equations for calculating of f_c from the programmed code are as follows:

$$f_c = \frac{87.5}{87.5 - \text{CODE}} * 1 \text{ kHz} \tag{2.6}$$

with CODE: 0–63, and

$$f_c = \frac{262.5}{137.5 - \text{CODE}} * 1 \text{ kHz} \tag{2.7}$$

with CODE: 64–127. where CODE is the data on pins D0–D6 (0–127). D6 is the most significant bit (MSB) [25].

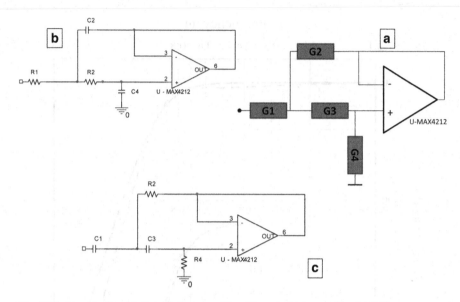

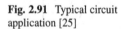

Fig. 2.90 Generic filter block (**a**), a low-pass filter (**b**), and a high-pass filter (**c**) [24]

Fig. 2.91 Typical circuit
application [25]

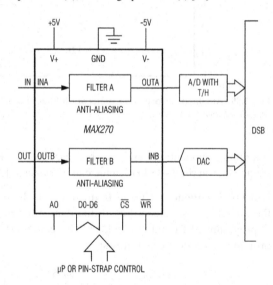

2.13   Memory Board

Modern microprocessors can manage a huge amount of data in short time; for that
they need that data are processed to the same speed.

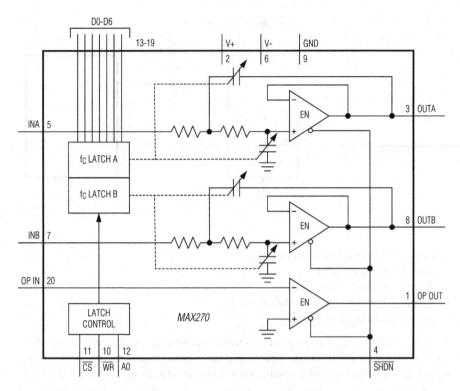

Fig. 2.92 MAX270 block diagram [25]

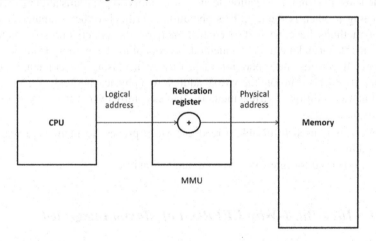

Fig. 2.93 Memory management

Memory is a large array of bytes, each with its own address. CPU fetches instructions that may cause loading from and storing to specific memory address. A instruction-execution cycle can be composed of:

Fig. 2.94 DS2431, typical application [26]

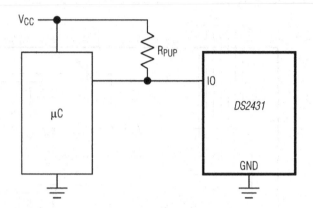

- Fetches an instruction from memory
- Decodes an instruction
- Fetches operands from memory
- Executes the instruction on the operands
- Stores results back in memory

Basic operation cycle of a computer is defined as instruction cycle, sometimes called fetch-and-execute cycle. This cycle is repeated continuously by the CPU, from bootup to when the computer is shut down. Memory management (Fig. 2.93) is the process of computer memory. The main requirement of memory management is to provide ways to dynamically allocate area of memory to programs at their request, and freeing it when is request. This operation is critical to the computer system. Several methods have been devised that increase the effectiveness of memory management. Virtual memory is a method of decoupling the memory from physical hardware. It permits the separation of processes increasing the efficiency of the memory. Its goal is translate the virtual address to a physical address.

In this way addition of virtual memory enables control over memory systems and methods of access:

Protection: It is used to disable a read and write process to memory that is not allocated to it.

Sharing: It is used for inter-process communication [6].

2.13.1   1024-Bit, 1-Wire EEPROM of Maxim Integrated

The DS2431 (Figs. 2.94 and 2.95) is a 1024-bit, 1-Wire EEPROM chip organized as four memory pages of 256 bits each. Data are written in an 8-byte scratchpad, verified, and then copied to the EEPROM memory. As a special feature, the four memory pages can individually be right-protected or put in EPROM-emulation mode, where bits can only be changed from a 1 to a 0 state. The DS2431

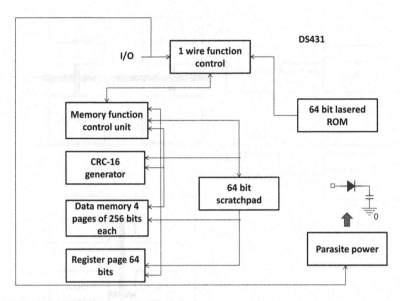

Fig. 2.95 DS2431, block diagram [26]

communicates over the single-conductor 1-Wire bus. The communication follows the standard 1-Wire protocol. Each device has its own unalterable and unique 64-bit ROM registration number that is factory lase-red into the chip. The registration number is used to address the device in a multidrop, 1-Wire net environment [26].

2.14 Bus Interface

A system bus is a device that connects several components of a electronic system. This technique was developed to reduce costs and improve modularity. The goal is based on management of data bus and address bus to determine what and where data must be sent; moreover, a control bus determine its operation.

According to the Figs. 2.96 and 2.97, data are moved from several electronic devices by means of buses.

Buses are designed in one of the following three ways:

a. Point to point connection: a particular bus is designed for every transfer.
b. Common bus: a common bus is used for all transfers.
c. Multiple bus: it is a combination of the previous two methods.

The goal of the bus is to connect CPU and memory; it represents the main characteristics of the system.

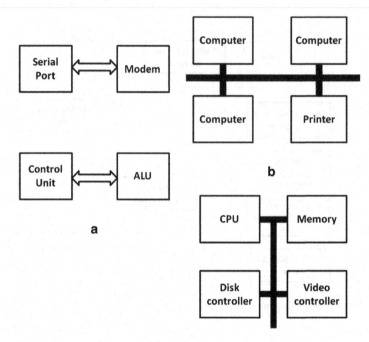

Fig. 2.96 Example of point-to-point buses (**a**) and multi-point buses (**b**)

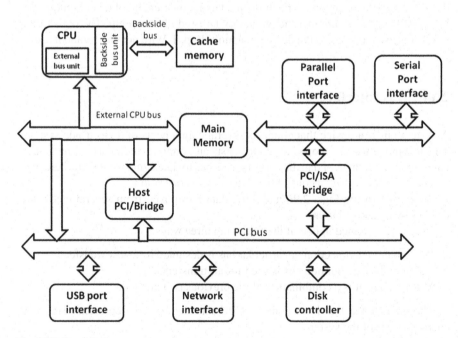

Fig. 2.97 Bus interface

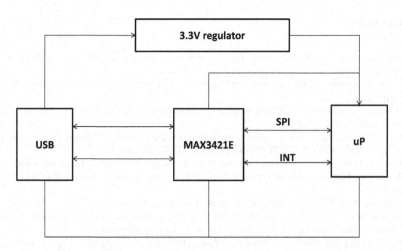

Fig. 2.98 MAX3421E, typical application [27]

Many CPUs use a set of pins for communicating with memory, but able to operate at different speeds with different protocols. Others used smart controllers to send the data directly in memory, and this is known as direct memory access. Moreover, most modern systems combine both solutions.

In modern systems high speed memory is built directly into the CPU, known as cache, using high-performance buses that operate at speed much greater than memory.

Such systems are architecturally more similar to multicomputers [6].

2.14.1 MAX3421E, USB Peripheral/Host Controller with SPI Interface

The MAX3421E (Fig. 2.98) USB peripheral/host controller contains the digital logic and analog circuitry necessary to implement a full-speed USB peripheral or a full-/low-speed host compliant to USB specification rev 2.0. An internal serial interface engine (SIE) handles low-level USB protocol details such as error checking and bus retries.

The MAX3421E makes the vast collection of USB peripherals available to any microprocessor, ASIC, or DSP when it operates as a USB host. For point-to-point solutions, for example, a USB keyboard or mouse interfaced to an embedded system, the firmware that operates the MAX3421E can be simple since only a targeted device is supported [27].

References

1. Park J, Mackqy S (2003) Pratical data acquisition for instrumentation and system control. Elsevier, Oxford
2. Lacanette K (1997) Temperature Sensor Handbook - Application Note National Semiconductor Corporation
3. National Instruments (2002) Data Acquisition Fundamentals, Application Note 007, National Instruments Corporation
4. National Instruments (1997) Signal conditioning fundamentals for PC-based data acquisition systems, Application Note 048, National Instruments Corporation
5. Taylor J (1986) Computer-Based Data Acquisition System - Instrument Society of America
6. Wikipedia - http://en.wikipedia.org/wiki/Analog-to-digital_converter - Wikimedia Foundation, Inc.
7. Vandoren A (1982) Data Acquisition System - Reston, Va. Reston Pub. Co.
8. Maxim Integrated (2012) Data sheet of DS1843 - Maxim Integrated. Copyright Maxim Integrated Products (http://www.maxim-ic.com). Used by permission
9. Maxim Integrated (2010) Understanding flash ADC - Maxim Integrated. Copyright Maxim Integrated Products (http://www.maxim-ic.com). Used by permission
10. Maxim Integrated (1996) Data sheet of MAX1150 - Maxim Integrated. Copyright Maxim Integrated Products (http://www.maxim-ic.com). Used by permission
11. Maxim Intergrated (2001) Deglitching techniques for high-voltage R-2R DACs - Maxim Integrated. Copyright Maxim Integrated Products (http://www.maxim-ic.com). Used by permission
12. Maxim Integrated (2005) Data sheet of MAX5109 - Maxim Integrated. Copyright Maxim Integrated Products (http://www.maxim-ic.com). Used by permission
13. Maxim Integrated (2008) Data sheet of MAX5893 - Maxim Integrated. Copyright Maxim Integrated Products (http://www.maxim-ic.com). Used by permission
14. Maxim Integrated (2011) Data sheet of MAXQ612/MAXQ622 - Maxim Integrated. Copyright Maxim Integrated Products (http://www.maxim-ic.com). Used by permission
15. Linear Tech. Corp. (1992) Data sheet of LT1128 - Linear Tech. Corp. Used by permission
16. Maxim Integrated (2011) Data sheet of MAX9632 - Maxim Integrated. Copyright Maxim Integrated Products (http://www.maxim-ic.com). Used by permission
17. Maxim Integrated (2012) Data sheet of MAX4638/4639 - Maxim Integrated. Copyright Maxim Integrated Products (http://www.maxim-ic.com). Used by permission
18. Maxim Integrated (2001) Using power management with high-speed microcontrollers - Maxim Integrated. Copyright Maxim Integrated Products (http://www.maxim-ic.com). Used by permission
19. Maxim Integrated (2011) Energy management for small portable systems - Maxim Integrated. Copyright Maxim Integrated Products (http://www.maxim-ic.com). Used by permission
20. Maxim Integrated (2011) Data sheet of MAX16920 - Maxim Integrated. Copyright Maxim Integrated Products (http://www.maxim-ic.com). Used by permission
21. Maxim Integrated (2010) Data sheet of MAX8662/MAX8663 - Maxim Integrated. Copyright Maxim Integrated Products (http://www.maxim-ic.com). Used by permission
22. Rabaye JM, Chandrakasan A, Nikolic B (2003) Digital Integrated Circuits 2nd - Prentice Hall
23. Maxim Integrated (2001) Data sheet of MAX9155 - Maxim Integrated. Copyright Maxim Integrated Products (http://www.maxim-ic.com). Used by permission
24. Maxim Integrated (2003) Tutorial, analog filter design demystified - Maxim Integrated. Copyright Maxim Integrated Products (http://www.maxim-ic.com). Used by permission

25. Maxim Integrated (2012) Data sheet of MAX270/MAX271 - Maxim Integrated. Copyright Maxim Integrated Products (http://www.maxim-ic.com). Used by permission
26. Maxim Integrated (2012) Data sheet of DS2431 - Maxim Integrated. Copyright Maxim Integrated Products (http://www.maxim-ic.com). Used by permission
27. Maxim Integrated (2007) MAX3421E, USB peripheral/host controller with SPI interface - Maxim Integrated. Copyright Maxim Integrated Products (http://www.maxim-ic.com). Used by permission

Chapter 3
Communication Bus

Abstract A bus is a system used to transfer data between components inside a electronic device, in particular a computer. Modern devices can use parallel, USB and serial connections for example, and can wired in either a multi-drop or daisy chain topology.

3.1 Bus USB and FireWire

A USB system consists in a asymmetric design with host controller and multiple daisy-chained devices. It was designed to avoid use of plug expansion cards into computer's PCI bus. For many devices such as digital cameras, USB has become the standard connection. Today USB is the most successful communication interface in use except for the monitors and high-quality digital video where require a higher data rate than USB as FireWire; even in the next years this limit will be solved.

3.1.1 Standardization and Technical Details

The design of USB (Figs. 3.1 and 3.2) is standardized by the USB Implementers Forum (USB-IF), an industry standards body incorporating leading companies from the computer and electronics industries.

A physical USB device may consist of several logical sub-devices; its communication is based on pipes (logical channels) as visualized in Fig. 3.3. There are two types of pipes: stream and message pipes. Message pipes are used in bidirectional applications, for indicating status response. Stream pipes, instead, are used in unidirectional applications that transfer data using an isochronous, interrupt, or bulk transfer [3]:

- Control transfers
- Isochronous transfers

M. Di Paolo Emilio, *Data Acquisition Systems: From Fundamentals to Applied Design*,   81
DOI 10.1007/978-1-4614-4214-1_3, © Springer Science+Business Media New York 2013

Pin	Name	Description
1	Vcc	+5 VDC
2	D-	Data-
3	D+	Data+
4	GND	Ground

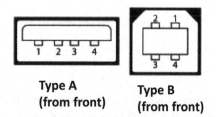

**Type A
(from front)**

**Type B
(from front)**

Fig. 3.1 Physical interface (USB)

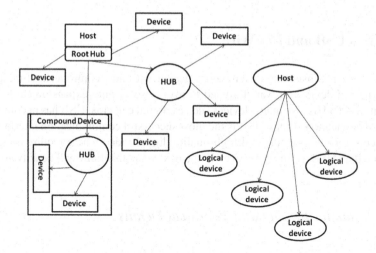

Fig. 3.2 Physical/logical bus topology (USB)

- Interrupt transfers
- Bulk transfers

    USB can support, for example, values of data rates as below:

- 1.5 Mbit/s (used for human interface devices)
- 12 Mbit
- 480 Mbit/s

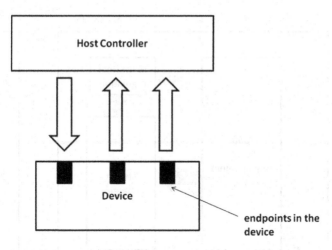

Fig. 3.3   Logical pipelines (USB)

3.1.2   USB Connectors

The connectors which the USB committee designed are support for the following features:

- The connectors are designed to be robust.
- The connectors are particularly cheap to manufacture.
- The connectors enforce the directed topology of a USB network.
- The USB standard specifies relatively low tolerances for compliant USB connectors, intending to minimize incompatibilities in connectors produced by different vendors [3].

3.1.3   Power Supply

USB has become as much a standard for connecting power to portable devices as it has for serial communication. Recently the power aspects of USB have been extended to cover battery charging as well as AC adapters and other power sources. A tangible benefit of this widespread use is the emergence of interchangeable plugs and adapters for charging and powering portable devices. This, in turn, allows charging from a far wider variety of sources than in the past when each device required a unique adapter. Arguably the most useful benefit of USB's power capabilities is the ability to charge batteries in portable devices. Nonetheless, there is more to battery charging than picking a power source, USB, or otherwise. This is particularly true for Li+ batteries, where improper charging cannot only shorten battery life, but also become a safety hazard. A well-designed charger optimizes

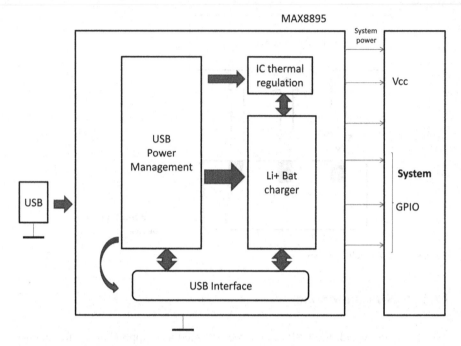

Fig. 3.4 MAX8895 [5]

safety and the user experience. It also lowers cost by reducing customer returns and warrantee repairs. Charging batteries with USB requires balancing battery "care and feeding" with the power limitations of USB and the size and cost barriers ever present in portable consumer device designs. The MAX8895 determines how best to use available input power without relying on the system to evaluate the power source. The available current can be used by the battery or the system, or it can be split between them. A built-in suspend timer automatically triggers suspend when no bus traffic is detected for 10 ms. In addition to automatically optimizing current from USB and adapter sources, the MAX8895 deftly handles switch-over from adapter and USB power to battery power; it allows the system to use all available input power when necessary (Fig. 3.4). This enables immediate operation with a dead or missing battery when power is applied. All power-steering MOSFETs are integrated, and no external diodes are needed. Die temperature is kept low by a thermal regulation loop that reduces charge current during temperature extremes [5].

3.1.4   USB Packet and Format

USB data are in the form of packets; all data are sent serially. Each USB data transfer consists of:

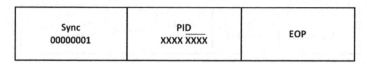

Fig. 3.5 Handshake packets (USB)

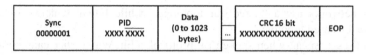

Fig. 3.6 Token packets (USB)

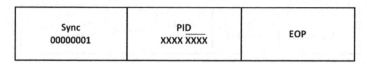

Fig. 3.7 Data packets (USB)

- A token packet
- An optional data packet
- A status packet

Moreover, USB packets may consist of the following fields:

1. Sync field: All the packets start with this sync field.
2. PID field: Identified the type of packet that is being sent.
3. ADDR field: Specified which device the packet is designated for.
4. ENDP field: This field is made up of 4 bits, allowing 16 possible endpoints.
5. CRC field: Cyclic redundancy checks are performed on the data within the packet payload.
6. EOP field: Indicates the end of packet.

The USB packets are defined in four basic types, each with a different format and CRC field:

1. Handshake packets, as visualized in Fig. 3.5.
2. Token packets, as visualized in Fig. 3.6.
3. Data packets, as visualized in Fig. 3.7. There are two basic data packets, DATA0 and DATA1.
4. PRE packet. Low-speed devices are supported with a special PID value, PRE.
5. Start of Frame Packets, as visualized in Fig. 3.8. Every 1 ms, the USB host transmits a special SOF (start of frame). This is used to synchronize isochronous data flows [3].

Sync 00000001	PID XXXX XXXX	Frame number	End point	CRC XXXXX	EOP

Fig. 3.8 Start of frame packets (USB)

3.1.5 FireWire

FireWire was designed as high-speed serial bus for audio and video equipment. The main differences between FireWire and USB can be described in the following points:

- USB networks use a "tiered-star topology," while FireWire networks use a "repeater-based topology."
- USB uses a "speak-when-spoken-to" protocol, a FireWire device can communicate with any other node at any time.
- A USB network relies on a single host at the top of the tree to control the network. In a FireWire network, any capable node can control the network.

In other words, USB was designed for simplicity and low cost, while FireWire was designed for high performance [1–3].

3.1.6 USB Peripheral Controller with SPI Interface, MAX3420E

The MAX3420E (Fig. 3.9) contains the digital logic and analog circuitry necessary to implement a full-speed USB peripheral compliant to USB specification rev 2.0. A built-in full-speed transceiver features ±15 kV ESD protection and programmable USB connect and disconnect. An internal serial-interface engine (SIE) handles low-level USB protocol details such as error checking and bus retries. The MAX3420E operates using a register set accessed by an SPI interface that operates up to 26 MHz. Any SPI master (microprocessor, ASIC, DSP, etc.) can add USB functionality using the simple 3- or 4-wire SPI interface.

Internal level translators allow the SPI interface to run at a system voltage between 1.71 and 3.6 V. USB timed operations are done inside the MAX3420E with interrupts provided at completion, so an SPI master does not need timers to meet USB timing requirements. The MAX3420E includes four general-purpose inputs and outputs, so any microprocessor that uses I/O pins to implement the SPI interface can reclaim the I/O pins and gain additional ones [6]

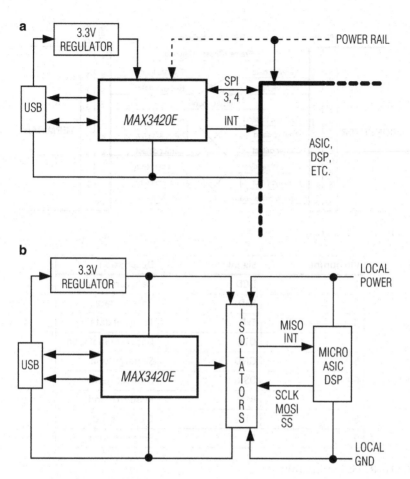

Fig. 3.9 MAX3420E: connected to a large chip (**a**), optical isolation of USB using the MAX3420E (**b**) [6]

3.2   Serial Communications

The RS-232 interface is the Electronic Industries Association (EIA) standard for the interchange of serial binary data between two devices. A typical RS-232 system (Figs. 3.10 and 3.11) is composed of two devices: data communicator (DCE) and Data terminal (DTE).

DTE stands for Data Terminal Equipment (e.g., a computer), and DCE stands for Data Communications Equipment (e.g., a modem). They are used to indicate the pin-out for the connection of the devices according to the direction of the signals.

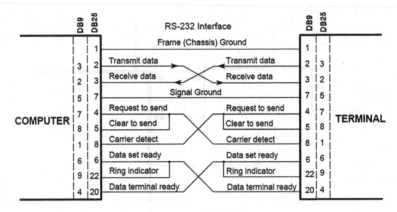

Fig. 3.10 RS232 interface

Pin number	Signal	Description
1	DCD	Data carrier detect
2	RxD	Receive data
3	TxD	Transmit data
4	DTR	Data terminal ready
5	GND	Signal ground
6	DSR	Data set ready
7	RTS	Ready to send
8	CTS	Clear to send
9	RI	Ring indicator

Fig. 3.11 RS232 interface-DB9

3.2.1 Signal Description

The signals of rs232 can be described as below according to Fig. 3.11:

TxD: This pin carries data from the computer to the serial device.

RXD: This pin carries data from the serial device to the computer.

DTR signals: DTR is used by the computer to signal that it is ready to communicate with the serial device like modem.

DSR: Similar to DTR, data set ready (DSR) is an indication from the dataset that it is ON.

DCD: Data carrier detect (DCD) indicates that carrier for the transmit data is ON.

RTS: This pin is used to request clearance to send data to a modem.

CTS: This pin is used by the serial device to acknowledge the computer's RTS Signal. In most situations, RTS and CTS are constantly on throughout the communication session. Clock signals (TC, RC, and XTC): The clock signals are only used for synchronous communications.

CD: CD stands for carrier detect. Carrier detect is used by a modem to signal that it
has made a connection with another modem, or has detected a carrier tone.
RI: RI stands for ring indicator.

3.2.2 Limitation of RS232

RS-232 has some serious limitations as electrical interface. First of all, serial
communications needs a common ground between DTE and DCE. This can be
clear in short cable, in an opposite way, with longer cables and connections between
devices can be a problem causing "uncommon grounds." Moreover, as the baud
rate and cable length increased, the effect of capacitance between cables introduces
crosstalk noise which can be reduced by using low capacitance. In the end, RS
232 was designed for communication of local devices (30–60 m maximum), and
supported one transmitter and one receiver [1].

3.2.3 MAX220-MAX249 for Serial Applications

The MAX220MAX249 family of line drivers/receivers is intended for all EIA/TIA-
232E and V.28/V.24 communications interfaces, particularly applications where
±12 V is not available. These parts are especially useful in battery-powered systems,
since their low-power shutdown mode reduces power dissipation to less than 5 μW.
The MAX220/MAX249 contain four sections: dual charge-pump DC–DC voltage
converters, RS-232 drivers, RS-232 receivers, and receiver and transmitter enable
control inputs. In Fig. 3.12 details of MAX220/MAX249 are shown [7].

3.3 Wireless, Ethernet, and Bluetooth

Ethernet communication (Fig. 3.13) consists of the division of a bytes stream into
shorter pieces called frames. Each frames contains data source, address, and error-
checking data that controls when the data are damaged and enable the device to
retransmit data source. The frame structure (Fig. 3.14) consists of the following
fields:

- Preamble: This allows the receiver's clock to be synchronized with the sender's.
- The Start Frame Delimiter: Indicates the start of frame.
- The Destination Address: Address where frame is sent.
- The source address: Address of data source.
- Length: Size of data in the ethernet frame.
- Data: The information sent by ethernet connection.

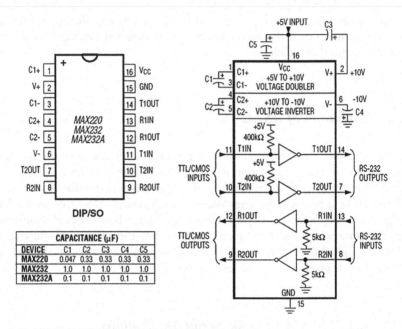

DIP/SO

CAPACITANCE (μF)					
DEVICE	C1	C2	C3	C4	C5
MAX220	0.047	0.33	0.33	0.33	0.33
MAX232	1.0	1.0	1.0	1.0	1.0
MAX232A	0.1	0.1	0.1	0.1	0.1

Fig. 3.12 MAX220-MAX249: typical application [7]

Fig. 3.13 Ethernet protocol

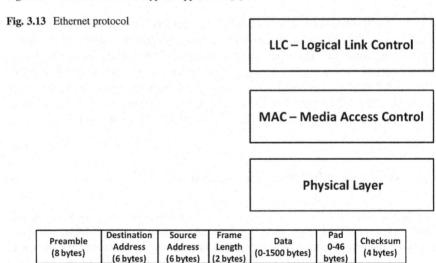

Preamble (8 bytes)	Destination Address (6 bytes)	Source Address (6 bytes)	Frame Length (2 bytes)	Data (0-1500 bytes)	Pad 0-46 bytes	Checksum (4 bytes)

Fig. 3.14 Ethernet frame

- Pad: To compensate the data transmission because frame must be at least 64 byte long.
- Checksum: Used for error detection.

Wireless telecommunication known as IEEE 802.11 is a set of standard for implementing the transfer of information between two points that are not physically connected. The frequency bands are 2.4, 3.6, and 5 GHz. Other examples of wireless technology are GPS, bluetooth, and so on.

The 802.11 family consists of a series of half-duplex over-the-air modulation techniques that use the same basic protocol. The most popular are those defined by the 802.11b and 802.11g protocols, which are correct to the original standard.

Current 802.11 standards define "frame" types for using in transmission; each frame consists of an MAC header, payload, and frame check sequence (FCS). The first two bytes of the MAC header form a frame control, and it is further subdivided into the following subfields:

- Protocol Version: The protocol version.
- Type: The type of WLAN frame.
- Sub Type: Addition discrimination between frames.
- ToDS and FromDS: They indicate whether a data frame is headed for a distribution system.
- More Fragments: The More Fragments bit is set when a packet is divided into multiple frames for transmission.
- Retry: Sometimes frames require retransmission, and for this there is a Retry bit which is set to one when a frame is resent.
- Power Management: This bit indicates the power management state of the sender after the completion of a frame exchange.
- More Data: The More Data bit is used to buffer frames received in a distributed system.
- WEP: The WEP bit is modified after processing a frame.
- Order: This bit is only set when the "strict ordering" delivery method is employed.

The next two bytes are reserved for the duration ID field. This field can take one of three following forms: Duration, Contention-Free Period (CFP), and Association ID (AID) [3].

Bluetooth (Fig. 3.15) is a method for data communication that uses short range. It is based on a nominal antenna power of 0 dbm. Each bluetooth stage has a free-running clock, CLKN, that determines timing and hopping of the transceivers [3,4].

3.4 GSM for Data Acquisition Systems

GSM (Global System for Mobile Communications, originally Groupe Spcial Mobile) is a communication standard developed by the European Telecommunications Standards Institute (ETSI) to describe protocols used by mobile phones. The network is composed of the following structures:

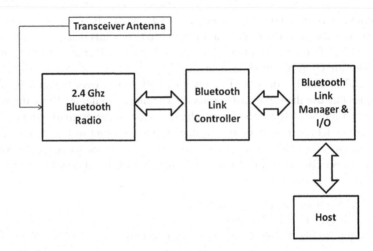

Fig. 3.15 Bluetooth

- The Base Station Subsystem (the base stations and their controllers)
- The Network and Switching Subsystem (the part of the network most similar to a fixed network)
- The GPRS Core Network (optional part)
- The Operations Support System (OSS) for maintenance of the network

GSM networks work in a different carrier frequency ranges, with most 2G GSM network that operates in the 900 Mhz or 1,800 Mhz bands.

Most 3G networks in Europe operate in the 2,100 MHz frequency band [3].

GSM uses a digital modulation format called 0.3 GMSK (Gaussian minimum shift keying). The 0.3 describes the bandwidth of the Gaussian filter with relation to the bit rate.

GMSK is a special type of digital FM modulation. 1s and 0s are represented by shifting the RF carrier by plus or minus 67.708 KHz. Modulation techniques which use two frequencies to represent one and zero are denoted as FSK (frequency shift keying). In the case of GSM the data rate of 270.833 kbit/s is chosen to be exactly four times the RF frequency shift. This has the effect of minimizing the modulation spectrum and improving channel efficiency. FSK modulation where the bit rate is exactly four times the frequency shift is called MSK (minimum shift keying). In GSM, the modulation spectrum is further reduced by applying a gaussian pre-modulation filter. This slows down the rapid frequency transitions which would otherwise spread energy into adjacent channels.

GSM uses TDMA (time division multiple access) and FDMA (frequency division multiple access). The frequencies are divided into two bands. The uplink is for mobile transmission and the down link is for base station transmission. Each band is divided into 200 KHz slots called ARFCN (Absolute radio frequency channel number).

Timing advance is required in GSM because it uses time division multiple access (TDMA). Since a radio signal can take a finite period of time to travel from the mobile to the base station, there must be some way to make sure the signal arrives at the base station at the correct time.

The GSM standards define a radio communications system that works properly only if each component part operates within precise limits. Essentially, mobiles and base stations must transmit enough power, with sufficient fidelity to maintain a call of acceptable quality, without transmitting excessive power into the frequency channels and time-slots allocated to others. Similarly, receivers must have adequate sensitivity and selectivity to acquire and demodulate a low level signal [9].

3.4.1 GPS Receiver, MAX2769

The MAX2769 is the industrys first global navigation satellite system (GNSS) receiver covering GPS, GLONASS, and Galileo navigation satellite systems on a single chip. This single-conversion, low-IF GNSS receiver is designed to provide high performance for a wide range of consumer applications, including mobile handsets. Designed on Maxims advanced, low-power SiGe BiCMOS process technology, the MAX2769 offers the highest performance and integration at a low cost. Incorporated on the chip is the complete receiver chain, including a dual-input LNA and mixer, followed by the image-rejected filter, PGA, VCO, fractional-N frequency synthesizer, crystal oscillator, and a multibit ADC. The total cascaded noise figure of this receiver is as low as 1.4 dB. In Fig. 3.16 a typical application circuit is shown [8].

3.5 PCI and PCI Express

The PCI architecture (Fig. 3.17) was designed as a replacement for the ISA standard, with three main features: increasing the performance during data transferring, to be independent, to simplify the system with the possibility of adding or removing peripherals. PCI architecture works with clock that run at 25 or 33 Mhz. Moreover, it is equipped with a 32-bit data bus and an expansion bus of 64-bit.

The PCI bus is the de-facto standard bus for current-generation personal computers and embedded applications. The main advantages of using embedded applications can be the following:

• Direct implementation in FPGAs
• Efficient protocol
• Ready availability of development hardware

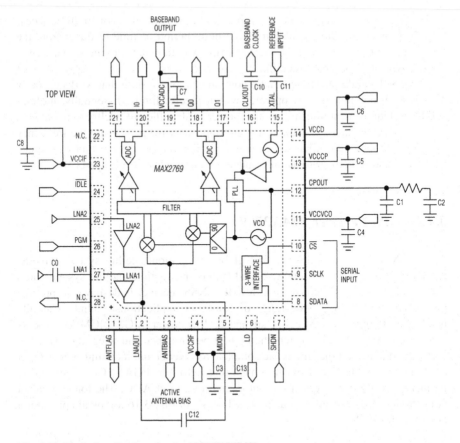

Fig. 3.16 Typical application circuit of MAX2769 [8]

In Figs. 3.18 and 3.19 typical read and write signals for PCI bus are shown.

An improvement of PCI is called PCI express (peripheral component interconnect express), officially abbreviated as PCIe. The main differences between PCI and PCI express are the following: more detailed error detection, smaller physical footprint, higher performance bus. In other words, PCIe can be seen as a high-speed serial replacement of the older PCI bus. Another important difference is the bus topology: PCIe uses a shared parallel bus architecture.

In terms of bus protocol, PCIe communication is encapsulated in packets.

Moreover, the stated goal of PCI express can be provided by [3]:

- A local bus for chip-to-chip interconnects
- A method to upgrade PCI slot performance at lower costs

Signal Name	Driven by	Description
CLK	Master	Bus Clock (normally 33MHz; DC okay)
FRAME#	Master	Indicates start of a bus cycle
AD[31:0]	Master/Target	Address/Data bus (multiplexed)
C/BE#[3:0]	Master	Bus command (address phase) Byte enables (data phases)
IRDY#	Master	Ready signal from master
TRDY#	Target	Ready signal from target
DEVSEL#	Target	Address recognized
RST#	Master	System Reset
PAR	Master/Target	Parity on AD, C/BE#
STOP#	Target	Request to stop transaction
IDSEL		Chip select during initialization transactions
PERR#	Receiver	Parity Error
SERR#	Any	Catestrophic system error

Fig. 3.17 Bus PCI-signals

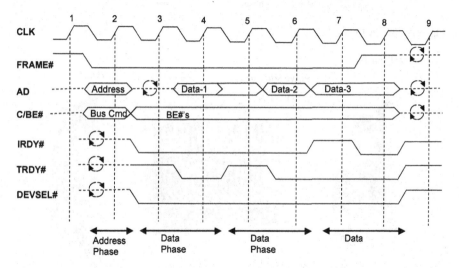

Fig. 3.18 Bus PCI-read

3.6   Standard VME

The VME bus (Versa Module Europa) is a flexible open-ended bus system based on the Eurocard standard (Figs. 3.20 and 3.21). It was introduced by Motorola, Phillips, Thompson, and Mostek in 1981. VME bus was intended to be a flexible environment supporting a variety of computing intensive tasks, and is now a very popular protocol in the computer industry. It is defined by the IEEE 1014–1987 standard. The system is modular and follows the Eurocard standard. VME card cages contain 21 slots, the first of which must be used as a crate manager or system controller.

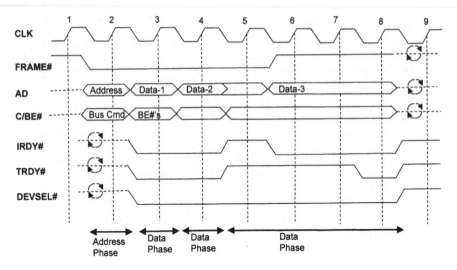

Fig. 3.19 Bus PCI-write

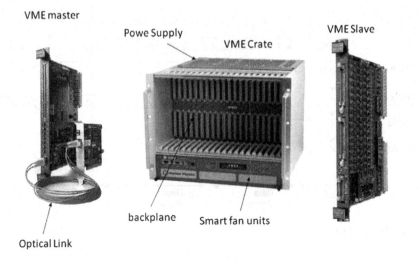

Fig. 3.20 Bus VME

A VMEbus (Fig. 3.21) is composed of four main buses:

Data Transfer: The data transfer bus consists of data lines (D31—D00), addressing lines (A31—A01, AM0—AM5, LWORD, DS0, and DS1), and control lines (AS*, WRITE*, DTACK, BERR, RETRY, DS0, and DS1*). Some data transfer bus lines are used for more than one purpose. Addressing lines are driven by a bus master and monitored by a bus slave. The master drives AS* (Address Strobe) to indicate a valid address on the bus. Data lines are used to transfer information across the bus. All data transfers are terminated by asserting one of the following signals: DTACK*

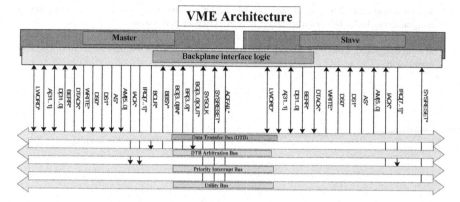

Fig. 3.21 Bus VME signals

(a data transfer acknowledgment indicating a successful transfer), BERR* (a Bus Error indicating a bus timeout or an error), or VME64 RETRY*.

Priority Interrupt Bus: the priority interrupt bus consists of interrupt request lines (IRQ7—IRQ1), the IACK* line, and IACK* daisy-chain lines (IACKIN* and IACKOUT*). This bus is used to assert an interrupt. It is also used in conjunction with the data transfer bus to acknowledge an interrupt.

Data Transfer Arbitration Bus: the data transfer arbitration bus is used by a bus master to request permission from the System Controller to use the bus. The arbiter in the System Controller determines which master will be granted the bus.

Utility Bus: the utility bus contains all system monitoring lines, including SYS-RESET*, BERR*, SYSFAIL*, and ACFAIL*. SYSRESET* is used to reset the VMEbus system. BERR* is used to terminate a bus cycle. SYSFAIL* is a utility signal that can be used for system diagnostics. ACFAIL* is asserted when the AC line voltage stops.

References

1. Park J, Mackqy S (2003) Pratical data acquisition for instrumentation and system control. Elsevier, Oxford
2. Taylor J (1986) Computer-Based Data Acquisition System - Instrument Society of America
3. Wikipedia -http://en.wikipedia.org/wiki/Usb- Wikimedia Foundation, Inc.
4. Vandoren A (1982) Data Acquisition System - Reston, Va. Reston Pub. Co.
5. Maxim Integrated (2010) The basics of USB Battery charging: a survival guide - Maxim Integrated. Copyright Maxim Integrated Products (http://www.maxim-ic.com). Used by permission
6. Maxim Integrated (2010) USB peripheral controller with SPI interface, MAX3420E - Maxim Integrated. Copyright Maxim Integrated Products (http://www.maxim-ic.com). Used by permission
7. Maxim Integrated (2009) Data sheet of MAX220MAX249 - Maxim Integrated. Copyright Maxim Integrated Products (http://www.maxim-ic.com). Used by permission

8. Maxim Integrated (2010) Data sheet of MAX2769 - Maxim Integrated. Copyright Maxim Integrated Products (http://www.maxim-ic.com). Used by permission
9. Maxim Integrated (2003) Introduction to GSM and GSM mobile RF transceiver derivation - Maxim Integrated. Copyright Maxim Integrated Products (http://www.maxim-ic.com). Used by permission

Chapter 4
Design of Data Acquisition Systems

Abstract In this chapter the main aspects of the design and a possible system will be described. The idea is to develop a design approach that can be a guide for future electronic designer. By this, an approach with general description of requirements will be presented.

4.1 Introduction to the Design

Today, computer-based system used in many applications are composed of simple stand-alone personal computer with board connected or microprocessor-based systems to realize a complete network of minicomputers. Such systems are used in many real time applications including process control and monitoring. Design process of data acquisition systems (DASs) is an obscure process. It is possible to consider system design in two principal phases: functional design and final design. The functional design is visualized in Fig. 4.1. The first step is to value the requirements according to the user. The second step is to translate the specific requirements in electric criteria, for example: accuracy and bandwidth. At the end, the third step is to establish a specific configuration of the system [1, 4].

4.2 Functional Design of High Speed Computer-Based DAS

The goal of this study is to describe a technical note used to design a high speed computer-based DAS [1–4].

M. Di Paolo Emilio, *Data Acquisition Systems: From Fundamentals to Applied Design*, 99
DOI 10.1007/978-1-4614-4214-1_4, © Springer Science+Business Media New York 2013

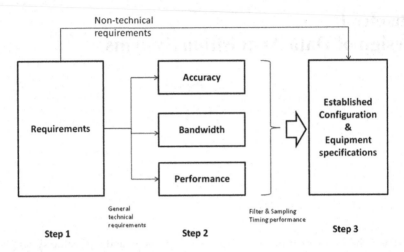

Fig. 4.1 Functional design: outline

4.2.1 Requirements

The requirements, in according to the first step, can be described in the following points (we will make an example):

Analog Input Channels: the system is designed to store 64 channels of analog data. The analog signals have a maximum differential voltage range of 10 V. Some transducers will be used to convert the signal in electrical form.

Analog Output Channels: the system is designed for a parallel transmission of eight analog signals reconstructed from digital form stored in memory board. All output analog signals (maximum voltage range 10 V–10 kΩ) are in singled-ended mode. According to the parallel transmission, the system is planned for simultaneous outputting of eight channels with bandwidth of DC-5 kHz.

Accuracy: the mean squared error between an analog signal and its reconstruction in output can be defined by the following equation:

$$\frac{\int_{T_1}^{T_2}[y(t)-x(t)]^2\mathrm{d}t}{\int_{T_1}^{T_2}[y(t)]^2\mathrm{d}t} \tag{4.1}$$

With $x(t)$ is the analog input signal, $y(t)$ is the analog output signal, and $T_2 - T_1$ is the analysis interval. This value must not exceed 0.2 %. This equation and corresponding variables are subject to the following conditions:

- Output signal are reconstructed from digital data stored in memory board.
- All input signals are considered of low pass type.
- The accuracy will be applied to each input and output signals at a data rate up to maximum frequency.
- System gain is unitary.

Analog Data Sampling Frequency: It is important to have an high accurate repro-
duction of the signals for each input corresponded to the bandwidth. The sampling
frequency can be selected manually or via computer. The system will permit the
simultaneous collection of 64 channels with bandwidth DC-5 kHz. The frequency
sampling is designed according to the highest frequency of the bandwidth, the
maximum level of attenuation for aliasing frequencies, and roll-off of the anti-
aliasing filter.

Anti-Aliasing Filters: Each analog input will be provided with an anti-aliasing filter;
it will be managed manually or via computer.

Minimum and Maximum Analysis Times: the system will be capable to store 64
analog channels with time interval from 10 ms to 1 s according to the sampling
frequency. Accuracy of time interval should be of about $\pm 25\,\mu$s.

Analysis Bandwidth: The data bandwidth of any input or output channel will
be operated from a minimum of DC-2.5 kHz to the maximum of DC-25 kHz in
conformity with the sampling frequency.

System Calibration Program: A program is designed to verify that the DAS is
operated following the requirements. The program will locate which element is not
in work.

Data Collection Program: A program that starts and stops the recording of 64 analog
channels. In particular the program is designed to have the following operations: set
the number of data input, select the bandwidth, the data interval, the sampling rate,
and the gain of input channel signal conditioner.

Transducer Calibration Program: A program is designed to make the test and
calibration of the transducers. All calibration data are stored in memory board or
exported in other mass storage devices.

Temperature: The system is designed to operate in a temperature range of, for
example, $-10°-+70°$ without damage. Normally, the system will be operated at
a temperature of 20°.

Shock and Vibration: The system is designed to resist from the repeated shock
and/or vibration. Vibrations of, for example, 15.24 mm double amplitude at frequency from 5 to 60 Hz.

4.2.2   Analysis of Accuracy (Static)

Equation (4.1) can be indicated in the following form:

$$\frac{\int_{T_1}^{T_2} [y(t) - x(t)]^2 \, dt}{\int_{T_1}^{T_2} [y(t)]^2 \, dt} \le \frac{0.2}{100} \tag{4.2}$$

$$E = y(t) - x(t) \tag{4.3}$$

$$y(t) = C \tag{4.4}$$

Then

$$\frac{\int_{T_1}^{T_2} [E]^2 dt}{\int_{T_1}^{T_2} [C]^2 dt} \leq 0.002 \tag{4.5}$$

$$\frac{E^2(T_2 - T_1)}{C^2(T_2 - T_1)} \leq 0.002 \tag{4.6}$$

$$E \leq (0.044)C \tag{4.7}$$

The static accuracy requirement is interpreted to be 0.044 times of input. If we design a data system with:

$$(0.044)C = 0.2 \tag{4.8}$$

and $I = 10\,\mathrm{V}$ for example, C will be 0.45 V.

4.2.3 Analysis of Accuracy (Dynamic)

About the accuracy of dynamic signal analysis, some possible discussion can be described as below:

Filter Pass-Band Ripple: We consider specifications of 0.1 dB for pass-band ripple. For full scale (10 V), this value is corresponded to an error of:

$$0.1\ \mathrm{dB} = 20\log(e_0/e_1), e_1 = 10\,\mathrm{V} \tag{4.9}$$

$$e_0 = 9.885 \tag{4.10}$$

The error is 1.2 %. At the level of 1 V, for example, the component of ripple error represents a value of 11.4 % on the reading.

Aperture and Input Filter Consideration: Aperture time is the width of the sampling window. The value of aperture time can be estimated assuming a sinusoidal input and calculating by the time required for input to change less than the resolution. For 14 bit converter with 5 kHz input, the time to maintain error less than resolution is 3.9 ns.

In the first chapter we have described the problems about the communication cabling; a cable for connecting the test transducer of the conditioner device is used. It is composed of resistance and capacitance, the capacitance in conjunction with the resistance of the transducer forms a low-pass RC filter that can attenuate high frequencies. Using a cable with about 160 pF/m (C) and output sensor impedance of 300 ohm (R), the effective filter for 100 m for example is:

$$RC = 300 * 160 * 10^{-12} * 100 = 3 * 10^{-6} s \tag{4.11}$$

The filter's transfer function can be described by the following equation:

$$G(i\omega) = \frac{1}{1+i\omega RC} \tag{4.12}$$

with

$$\text{Amplitude}: |G(i\omega)| = \frac{1}{(1+(\omega RC)^2)^{1/2}} \tag{4.13}$$

$$\text{Phase}: \theta = -\tan^{-1}(\omega RC) \tag{4.14}$$

The effective filter for the cable and sensor has a cutoff of 53 kHz. According to the specifications, we want to process the information for all sensors at frequencies up to 5 and 25 kHz. The errors introduced can be valued in the following way: Amplitude 0.44 % and phase $-5.4°$ at 5 kHz, Amplitude 9.5 % and phase $-25°$ at 25 kHz. The sampled data are used to reconstruct the analog signal by using a D/A converter, then this value is compared with input to value the error. The error is described by a zero-order hold:

$$\text{AmplitudeError}, \% = (1-\sin x/x) \times 100 \tag{4.15}$$

with

$$x = \pi(f/f_s) \tag{4.16}$$

f_s is the sampled frequency; and for the phase:

$$\text{PhaseDegrees} = \pi(f/f_s) \tag{4.17}$$

ADC Resolution According to magnitude of the error, we can estimate the resolution of ADC considering that all aliases must be attenuated by at least 80 dB, thus implying that the system's dynamic range is at least 80 dB:

$$db = 20\log(\text{ADC} - \text{resolution}) = 0.0001 \tag{4.18}$$

Thus, it is required a 14-bit A/D converter.
Bandwidth The bandwidth is calculated to be 5 kHz expandable to 25 kHz for some channels conveniently selected, with all aliases attenuated by at least 80 dB.

We assume that the frequency content of the analog signal is visualized in Fig. 4.2, with f_c the highest frequency of interest. If we use a filter with cut-off f_a with roll-off rate Rdb/octave, we can conservatively establish the highest frequency as f_a. We can choose the holding frequency f_n to be:

$$\log f_n = \log f_c + \frac{\log f_a - \log f_c}{2} = \frac{1}{2}\left(\log f_c + \log f_a\right) \tag{4.19}$$

$$f_n = \log^{-1}\left(\frac{1}{2}\log f_a\right) \tag{4.20}$$

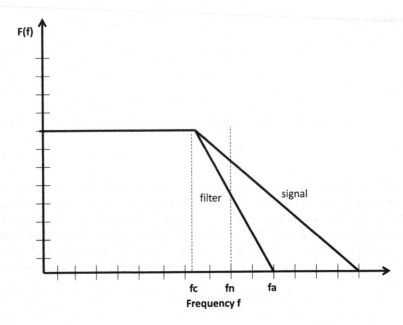

Fig. 4.2 Frequency content

Rolloff Rate	N	F_c (kHz)	F_n (kHz)	F_s (kHz)
12 dB/octave	6.6	500	50	100
24	3.3	50	15.8	31.6
36	2.2	23.2	10.8	21.5
48	1.6	15.8	8.9	17.8
80	1	10	7.1	14.1

Fig. 4.3 Filtering and sampling rate

To establish an acceptable distortion we can value it by using the following equation:
Establish number of octaves above f_c where input is attenuated by dB level:

$$N = \frac{\text{Attenuation} - \text{dB}}{RdB/\text{octave}} \tag{4.21}$$

Folding and sampled frequency:

$$f_n = \log^{-1}\left(\frac{1}{2}\log f_a\right) \tag{4.22}$$

$$f_s \geq 2f_n \tag{4.23}$$

Considering a distortion of 80 dB we can value the filtering with consideration according to Fig. 4.3: Based on these data, the sampling rate for individual channel for a 5-kHz of bandwidth can change from 14 to 100 kHz depending of roll-off rate of the filter.

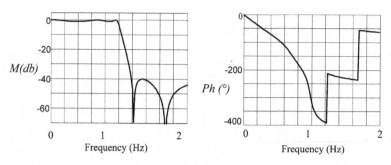

Fig. 4.4 Elliptic filter

Filter selection: Conforming with above considerations, we can use a elliptic filter (Fig. 4.4) defined by the following parameters:

- Roll off rate: 80 dB/octave
- Ripple: 0.1 dB pass band and stop band
- Phase linearity: linear over the range: $0.1 \leq \frac{f}{f_c} \leq 0.33$

An elliptic filter (also known as a Cauer filter) is a filter with equalized ripple behavior in both of the passband and the stopband. The ripple can be adjustable for each band independently. The gain of a lowpass elliptic filter is given by:

$$G_n(\omega) = \frac{1}{\sqrt{1 + \varepsilon^2 R_n^2(\zeta, \omega/\omega_0)}} \qquad (4.24)$$

where: ω_0 is cut-off frequency, ε is the ripple factor, ζ is the selectivity factor, and R_n is the nth-order of Chebyshev rational function.

Sampling Frequency: For a bandwidth of 5 kHz, we design the cut-off of the filter to 15 kHz to obtain linear phase requested. With $f_c = 15$ kHz, we compute the following parameters: N = 1, $f_a = 30$ kHz (effective cut-off frequency), $f_n = 21.2$ kHz, and $f_s = 42.4$ kHz. With these, amplitude error is 2.3 % and phase is 3.7°.

Signal Conditioner, Amplifier, Multiplexer, and ADC: Each channel required to include a differential amplifier which CMRR of 80 dB, a differential input impedance of 20 MΩ, offset current of 30 nA, and adjustable sensitivity from ±1 to ±100 mV. Instrumentation amplifiers (differential amplifier) amplify small differential voltages in the presence of large common-mode voltages, while offering a high input impedance. This characteristic has made them attractive to a variety of applications, such as strain-gauge bridge interfaces for pressure and temperature sensing, thermocouple temperature sensing, and a variety of low-side and high-side current-sensing applications [5]. The classic three-op-amp instrumentation amplifier (see Fig. 4.5) offers excellent common-mode rejection and accurate differential gain programmable by a single resistor. The architecture is based on a two-stage configuration: the first stage provides unity common-mode gain and all (or most) of the differential gain, while the second stage provides unity (or small) differential-

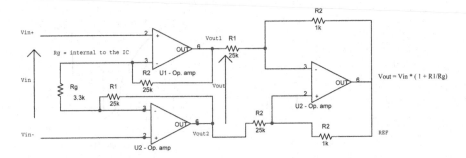

Fig. 4.5 Differential amplifier

mode gain and all of the common-mode rejection. Example of differential amplifier is MAX4198/99. The MAX4198/MAX4199 low-power, rail-to-rail differential amplifiers are ideal for single-supply applications that benefit from a low 0.01 % gain error. The MAX4198 is factory trimmed to a fixed gain of +1 V/V, and the MAX4199 is trimmed to a fixed gain of +10 V/V. Capable of operating from a single +2.7 to +7.5 V supply or from dual ±1.35 to ±3.75 V supplies, they consume only 42 μA while achieving −3 dB bandwidths of 175 kHz (MAX4198) and 45 kHz (MAX4199). These amplifiers feature a shutdown mode that reduces the supply current to 6.5 μAA. The MAX4198/MAX4199 can drive 5 kΩ loads to within 100 mV from each rail. The standard differential amplifier configurations provide common-mode rejection of 90 dB for the MAX4198 and 110 dB for the MAX4199. The input common-mode voltage range for the MAX4198 extends 100 mV Beyond-the-Rails [6].

The system is required to support 32 channels concurrently. It is preferable if a multiplexer/ADC is chosen with the sampling capability to accommodate all 32 channels, in other case we should use multiple ADC. We can define aggregate throughput F_s based on sampling frequency f_s for 5 kHz of bandwidth:

$$F_s = n - \text{channels} * f_s = 1357 \text{ kHz} \tag{4.25}$$

We can use eight separate 1 MHz ADCs-14 bit MAX1324 of Maxim, each of 32 input channels: the processor must be capable to accept 2 Mbyte/s of data at the same time.

The MAX1324 (Fig. 4.6) is 14-bit ADC eight channels independently selectable. Simultaneous sampling of all active channels preserves relative phase information. These devices are available ±10 V input ranges. The 0 to +5 V devices feature ±6 V fault-tolerant inputs. All eight channels convert in 3.8 μs, with a maximum eight-channel throughput of 263 ksps per channel. Internal or external reference and internal- or external-clock capability offer great flexibility and ease of use. A write-only configuration register can mask out unused channels, and a shutdown feature reduces power. A 16.6-MHz, 14-bit, parallel data bus outputs the conversion result [7].

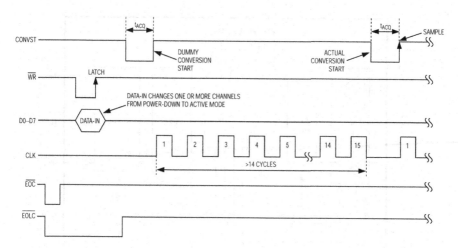

Fig. 4.6 Timing diagram of MAX1324 of Maxim [7]

Figure 4.7 shows the functional diagram of these devices. To preserve phase information across these multichannel devices, each input channel has a dedicated T/H amplifier. Use a low-input source impedance to minimize gain error harmonic distortion. The time required for the T/H to acquire an input signal depends on the input source impedance. If the input signals source impedance is high, the acquisition time lengthens and more time must be allowed between conversions. The acquisition time (t_1) is the maximum time that the device takes to acquire the signal. Use the following formula to calculate acquisition time:

$$t_1 = 10 * (R_S + R_{IN}) * 6pF \qquad (4.26)$$

where $R_{IN} = 2.2\,\text{k}\Omega$, R_S is the input signals source impedance, and t_1 is never less than 180 ns. A source impedance of less than $100\,\Omega$ does not significantly affect the ADCs performance. To improve the input-signal bandwidth under AC conditions, drive the input with a wide band buffer ($>50\,\text{MHz}$) that can drive the ADCs input capacitance and settle quickly. The T/H aperture delay is typically 13 ns. The aperture delay mismatch between T/Hs of 50 ps allows the relative phase information of up to eight different inputs to be preserved. During shutdown, the analog and digital circuits in the device power down and the device draws less than $100\,\mu\text{A}$ from AVDD, and less than $100\,\mu\text{A}$ from DV_{DD}. Select shutdown mode using the SHDN input. Set SHDN high to enter shutdown mode. After coming out of shutdown, allow a 1-ms wake-up time before making the first conversion. When using an external clock, apply at least 20 clock cycles with CONVST high before making the first conversion. When using internal-clock mode, wait at least $2\,\mu\text{s}$ before making the first conversion [7]. The outline of the DAS can be visualized in Fig. 4.8.

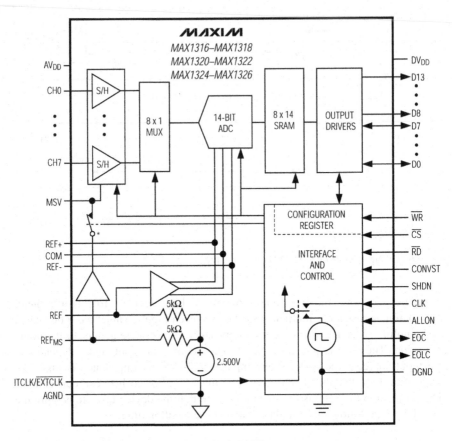

Fig. 4.7 Functional diagram of MAX1324 of Maxim [7]

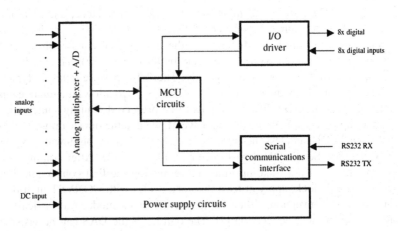

Fig. 4.8 Outline of DAQ hardware

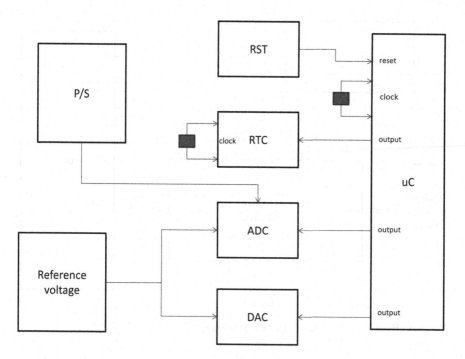

Fig. 4.9 Block diagram of portable data acquisition system [8]

4.3 Portable DAS

The hardware and software design of a portable measurement system is more complicated than just choosing the right IC to meet the required electrical performance. There are many tradeoffs to be considered. For example: size respect to the board, total cost, life cycle, and so on. At one extreme is a discrete design using readily available standard components and at the other extreme is a custom design using a single mixed-signal chip. The following will compare a discrete design to a design using the MAX1407 of Maxim Integrated.

Figure 4.9 shows the functional block diagram for a portable DAS using discrete components. The system includes power, analog, and digital circuitry. The power is derived from the batteries, with a power supply for a regulated output voltage. The battery voltage can be measured on demand with the A/D converter and there are reset circuits for the battery and the regulated output voltage. The analog includes the analog front end (AFE) circuitry (which may include D/A converters, op-amps, and analog switches) to interface with an electrochemical sensor, a thermistor circuit to measure temperature, an A/D converter with an input multiplexer, and a voltage reference for use with the AFE, thermistor, and A/D converter. The digital circuitry includes a 32-kHz oscillator, real-time clock (RTC), and a microcontroller including a high-frequency crystal, external interrupts for the user interface, internal memory,

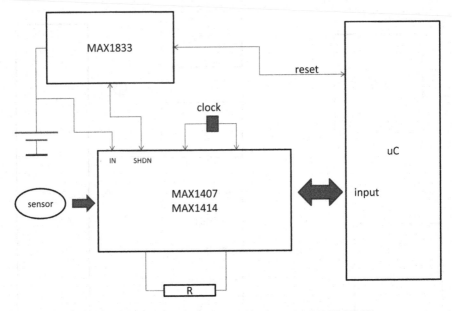

Fig. 4.10 Block diagram of portable data acquisition system using MAX1407 [8]

and possibly other peripherals, such as a liquid crystal display (LCD) bias and driver circuit to interface to an LCD. The sensor shown in Fig. 4.9 is an electrochemical type sensor configured in a counter configuration where the working electrode (WE) is biased via the FB1 pin. The external resistor between FB1 and OUT1 configures the force/sense D/A converter as a transimpedance amplifier to measure the current through the sensor. The reference electrode (RE) is biased via the FB2 pin, and the feedback loop is closed through the sensor impedance to the OUT2 pin to regulate a constant potential between WE and RE over varying sensor currents. In Fig. 4.10 is shown, instead, the functional block diagram for the same portable DAS using the MAX1407, simplifying the system design in Fig. 4.9. The MAX1407 contains all the circuitry required [8].

4.4   Design Guidelines for High-Performance, Multichannel

Many advanced industrial applications require the use of high-performance, simultaneous-sampling, multichannel ADCs. Consider as examples an advanced power-line monitoring (Fig. 4.11) or contemporary three-phase motor-control system (Fig. 4.12). These applications require accurate simultaneous, multichannel measurement over a wide dynamic range of 70–90 dB (depending on the application). A sample rate of 16 ksps or higher is common. Each power phase

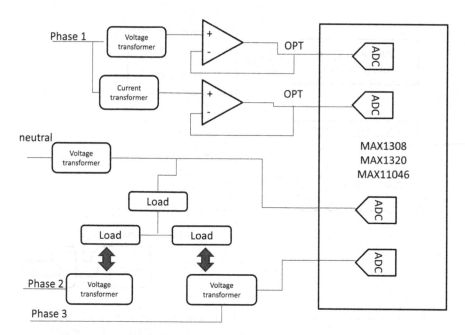

Fig. 4.11 Advanced power-line monitoring

in Fig. 4.11 is represented by a current transformer (CT) and a voltage transformer (PT). The complete system is comprised of four such pairs (one pair for each of the three phases plus neutral).

Two classes of noise/interference can be defined in the DAQ. The first class of noise comes from the internal electronic component noise. Sources include ADC conversion process noise and harmonic distortion, buffer amplifier noise and distortions, and reference noise and stability. A second source of interference is the system's external environment. Examples include external electromagnetic noise, power-supply noise/ripple, I/O pin crosstalk, and digital system noise and interference. These various noise sources are shown in Fig. 4.13.

General PCB Layout Guidelines: Several important PCB guidelines will help achieve the best performance in multichannel, simultaneous-sampling DAQ (DAQ or DAS) applications.

- Use PC boards with ground planes.
- Ensure that the analog and digital lines are separated from each other.
- Do not run digital and analog lines parallel to one another.
- Avoid running digital lines underneath the ADC package.
- Use a single, solid GND plane with digital signals routed from one direction and analog signals from the other.

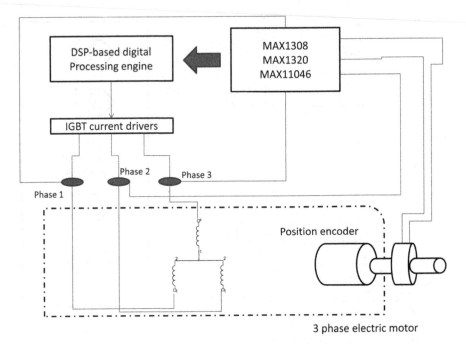

Fig. 4.12 Typical power-grid monitoring application [9]

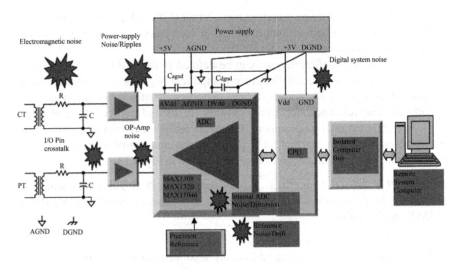

Fig. 4.13 Noise in a data acquisition system [9]

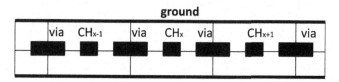

Fig. 4.14 Example of analog input routing [9]

- Keep the ground return to the power supply for this ground low impedance and as short as possible for noise-free operation.

Input PCB traces which carry sensitive analog signals from the connector to the ADC inputs can be subject to noise, interference, and channel-to-channel crosstalk. Special ground and signal shielding of these analog traces is critical to the integrity of the input signal. Figure 4.14 shows an example of a PCB layout intended to protect the analog signal [9].

References

1. Park J, Mackqy S (2003) Pratical data acquisition for instrumentation and system control. Elsevier, Oxford
2. Lacanette K (1997) Temperature Sensor Handbook - Application Note National Semiconductor Corporation
3. National Instruments (2002) Data Acquisition Fundamentals, Application Note 007, National Instruments Corporation
4. Taylor J (1986) Computer-Based Data Acquisition System - Instrument Society of America
5. Maxim Integrated (2007) There is a crowd for instrumentation amplifiers. Tutorial 4034. Maxim Integrated. Copyright Maxim Integrated Products. http://www.maxim-ic.com. Used by permission
6. Maxim Integrated (2004) Data sheet MAX4198/MAX4199. Maxim Integrated. Copyright Maxim Integrated Products. http://www.maxim-ic.com. Used by permission
7. Maxim Integrated (2004) Data sheet MAX1324. Maxim Integrated. Copyright Maxim Integrated Products. http://www.maxim-ic.com. Used by permission
8. Maxim Integrated (2001) The MAX1407 complete data acquisition system simplifies your system designs. Application note, Maxim Integrated. Copyright Maxim Integrated Products. http://www.maxim-ic.com. Used by permission
9. Maxim Integrated (2009) Design guidelines for high-performance, multichannel. Application note, Maxim Integrated. Copyright Maxim Integrated Products. http://www.maxim-ic.com. Used by permission

Chapter 5
Software for Data Acquisition Systems

Abstract In this chapter the main aspects of the software design for data acquisition (DAQ) systems are described. In particular we will analyze the rules to implement a good software control system to manage the DAQ systems.

5.1 Introduction

The design of data acquisition (DAQ) and control software must have the capacity to recover gracefully from instrument component failures and power outages without losing data. Data Acquisition software must be easily reconfigurable and provides a high level language for algorithm design. Moreover, it is required a data archiving method that can verify the integrity of the data acquired. In the software design, it is possible to confront with two points:

- For customer: a user-friendly software.
- For supplier: a powerful mechanism which can reflect all the combination of hardware facilities.

Good design is a negotiation of a process that tries to reconcile the two points, working top-down from the client side and bottom-up from the supplier side. The process of software design (Fig. 5.1) then consists of developing intermediate algorithms of abstraction: a set of efficiently implementable algorithms [1–4]. Software package usually has analysis and presentation capabilities into the DAQ software. Your application software normally does such tasks as:

- Real-time monitoring
- Data analysis
- Data logging
- Control algorithms
- Human machine interface (HMI)

M. Di Paolo Emilio, *Data Acquisition Systems: From Fundamentals to Applied Design*, 115
DOI 10.1007/978-1-4614-4214-1_5, © Springer Science+Business Media New York 2013

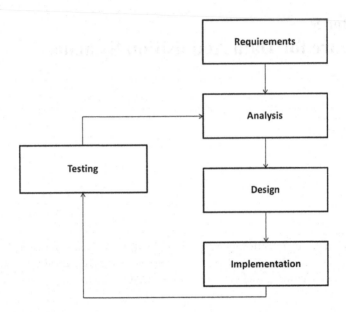

Fig. 5.1 Design software

5.2 LabView

LabView is a graphical programming language from National Instruments. It includes the Front Panel and the Block Diagram. Front Panel is like User Interface, can be composed of: controls inputs, shows outputs, and connects to the engine. Block Diagram, instead, allows it to function and connects everything together. Some advantages can be described in the following points [1–4]:

- Good for building piecewise: using small amounts of code in a larger code
- Visually programming is easier to learn
- NI has many pieces of hardware that are easily compatible with LabVIEW, but also can be connected to any hardware

5.3 Android for DAQ

Android is a Linux-based operating system designed primarily for smartphones and tablet computers. It is currently developed by Google in conjunction with the Open Handset Alliance. Android consists of a kernel based on the Linux kernel 2.6 and Linux Kernel 3.x (Android 4.0 onwards), with libraries and APIs written in C language and application software running on an framework which includes

Java-compatiblelibraries based on Apache Harmony. The main hardware platform for Android is the ARM architecture. It supports for x86 from the Android x86 project.

Personal computer gave engineers the ability to create smaller and more portable DAQ systems using C-language, for example, are tablet computers the natural evolution of this trend?

Tablets work under different operating systems and have less processing power than PC.

There are two points to consider: C and Labview are not supported in Android or IOS; USB, a common DAQ bus, is found in a set of Tablets, in other case, an adapter is required [1–4].

In these instances, moving a possible data acquisition system to a tablet requires extra attention. The tablet offers development environment for DAQ application.

There are three main considerations in creating a DAQ application on a tablet:

- Hardware connectivity: the Tablet has few options of control, such as Wi-fi and bluetooth.
- Program language support: Tablets support Android done in Java.
- Device driver availability: device drivers permits a high-level mode to easily and reliably execute DAQ board functionality.

5.4 Design of Firmware

Firmware is a software program or set of instructions programmed on a hardware, installed on ROM chips (ROM, PROM, EPROM) or flash chips. It enables the device to render its capabilities functionally. Moreover, it coordinates the activities of the hardware during normal operation and contains programming constructs used to perform those operations. Example: in a typical modem, the firmware will be a factor in establishing the modem's data rate, command set recognition, and special feature implementation. The node firmware (Fig. 5.2) can be seen as being formed from three entities:

1. Data structures that hold state information, behavior information, and map information.
2. Background processes that handle actions implemented in hardware or through the use of hardware peripherals such as the wireless communication and the sensor readings.
3. The main firmware code that executes the node's activity. The flowchart in Fig. 5.2 describes the basic operations if the node is in the idle state. The node can also be in the attached and the sensing state, respectively [1–4].

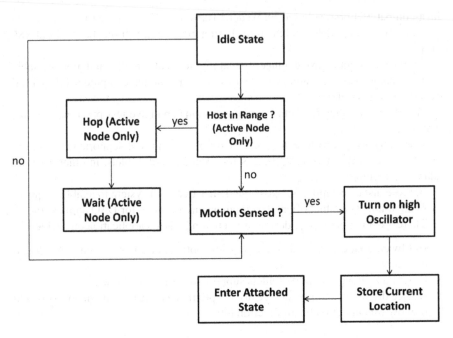

Fig. 5.2 Example of flow diagram of firmware

5.5 Example of Implementation of a Software for Data Acquisition System via VME Bus

The idea is to realize a software using the C language for the management of Versa Module Europa (VME)-based DAQ systems. The VME bus is a flexible open-ended bus system based on the Eurocard standard. The hardware is designed according to the last chapter adding interface for communication via VME bus. The structure of the software can be implemented for other DAQ system according to the bus interface used; if a USB interface is used it is necessary to implement C libraries (or another programming language) for USB communications. The software can also be implemented using LabView and its correspondents VME libraries; moreover, similar studies can be done for USB bus using appropriate libraries. The idea is to realize, starting from the VME universe drivers and libraries for the VME bus and the standard C libraries, a new set of functions and structures that assures the easy management of VME-based DAQ systems (Fig. 5.3). The system architecture relies on a VME crate managed by an Intel-based crate controller running the Linux operating system. The software (Figs. 5.4 and 5.5) is composed of five subsystems, each having a specific function:

- VME bus interface: implements the communication with the boards (DAQ hardware).

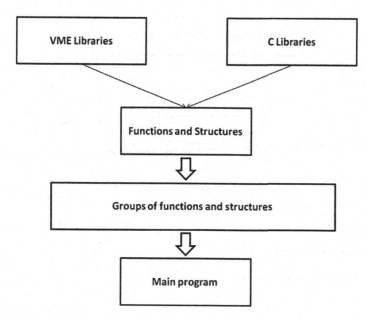

Fig. 5.3 Software structure

- Data writing: takes care of writing acquired data on structured data files.
- Configuration file interpreter: reads and parses the configuration file and sets up the DAQ.
- Error handler: manages errors that may show up during data taking (network problems, VME bus errors, disk access problems, ...)
- Network data transfer manager: takes care of transferring acquired data from the DAQ hardware to an optional data storage host via an Ethernet connection.

It has two user interfaces: a text-based user interface (TUI) and a graphical user interface (GUI) designed available by means of object-oriented language (C++). Both interfaces permit DAQ management with full acquisition control.

The configuration files configure the initial settings for DAQ board. Usually, it is used for user applications, server processes, and operating system settings.

The GUI is a type of user interface that allows users to interact with electronic devices using images rather than text commands. The use of GUI does not require knowledge of program code and structure of configuration file. A GUI uses a combination of technologies and devices to provide a platform that the user can interact with, for the tasks of gathering and producing information. Using GUI or TUI it is possible to define DAQ setup and all functionally needs to control it [1–6].

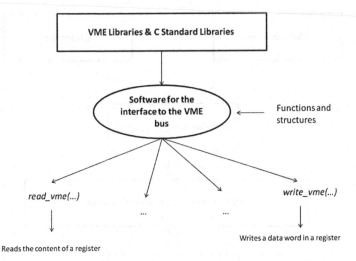

Fig. 5.4 Software structure: functions

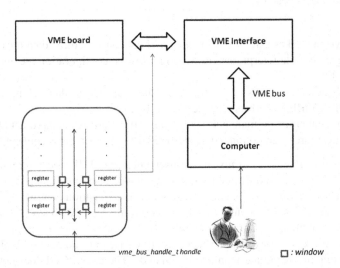

Fig. 5.5 VME bus, software interface

References

1. Park J, Mackqy S (2003) Pratical data acquisition for instrumentation and system control. Elsevier, Oxford
2. Lacanette K (1997) Temperature Sensor Handbook - Application Note National Semiconductor Corporation
3. National Instruments (2002) Data Acquisition Fundamentals, Application Note 007, National Instruments Corporation
4. Wikipedia -http://en.wikipedia.org/wiki/Firmware- Wikimedia Foundation, Inc.
5. Di Paolo Emilio M, Stalio S (2006) V-DAS (A versatile data acquisition software) The user interfaces. INFN Laboratori Nazional del Gran Sasso LNGS/TC-01-2006
6. Di Paolo Emilio M, Stalio S (2006) V-DAS (A versatile data acquisition software): the interface to the VME bus and the configuration file interpreter. INFN Laboratori Nazional del Gran Sasso LNGS/TC-02-2006

Chapter 6
Smart Data Acquisition System

Abstract In this chapter we describe a possible smart data acquisition system with the MAX1329, Maxim Integrated component using all concepts described in the last chapters.

6.1 General Description of MAX1329

The MAX1329/MAX1330 (Figs. 6.1–6.3) are smart data acquisition systems (DASs or DAQs) based on a successive approximation register (SAR) analog-to-digital converter (ADC). These devices are highly integrated, offering an ADC, digital-to-analog converters (DACs), operational amplifiers (op amps), voltage reference, temperature sensors, and analog switches in the same device.

The MAX1329 (Figs. 6.4 and 6.5) offers a single ADC with a reference buffer. The ADC is capable of operating in one of two user-programmable modes. In normal mode, the ADC provides up to 12 bits of resolution at 312 ksps. In DSP mode, the ADC provides up to 16 bits of resolution at 1,000 sps. The ADC accepts one external differential input or two external single-ended inputs as well as inputs from other circuitry on-board. An on-chip programmable gain amplifier (PGA) follows the analog inputs, reducing external circuitry requirements. The PGA gain is adjustable from 1 to 8 V/V. The MAX1329 operate from a 1.8 to 3.6 V digital power supply. Shutdown and sleep modes are available for power-saving applications. Under normal operation, an internal charge pump boosts the supply voltage for the analog circuitry when the supply is <2.7 V. The MAX1329 offer four analog programmable I/Os (APIOs) and four digital programmable I/Os (DPIOs). The APIOs can be configured as general purpose logic inputs and outputs, as a wake-up function, or as a buffer and level shifter for the serial interface to communicate with slave devices powered by the analog supply, AVDD. The DPIOs can be configured as general purpose logic inputs and outputs as well as inputs to directly

M. Di Paolo Emilio, *Data Acquisition Systems: From Fundamentals to Applied Design*,
DOI 10.1007/978-1-4614-4214-1_6, © Springer Science+Business Media New York 2013

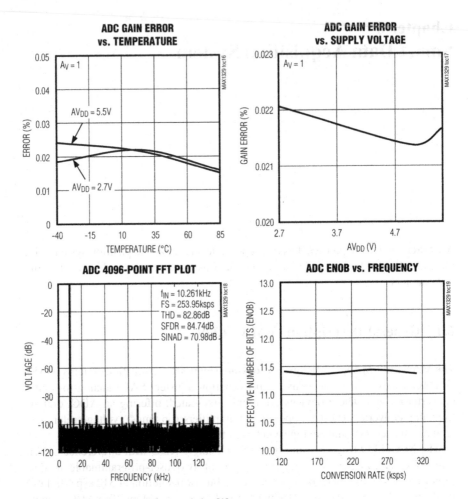

Fig. 6.1 MAX1329, typical characteristics [1]

control the ADC conversion rate, the analog switches, the loading of the DACs, wake-up, sleep, and shutdown modes, and as an interrupt when the analog-to-digital conversion is complete. The MAX1329 includes dual 12-bit force-sense DACs with a programmable reference buffer and one op amp. For the MAX1329, a 16-word DAC FIFO can be used with the DACA for direct digital synthesis (DDS) of waveforms [1].

Features:

- 1.8–3.6 V single digital supply operation
- Internal charge pump for analog circuits (2.7–5.5 V)

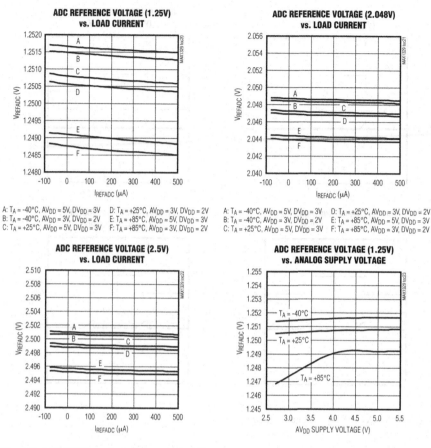

Fig. 6.2 MAX1329, typical characteristics [1]

- 12-Bit SAR ADC; 12 bits, 312 ksps, no missing codes; 16 bits, 1,000 sps, DSP mode; 16-word FIFO and 20-bit accumulator
- PGA with gains of 1, 2, 4, and 8
- Unipolar and bipolar modes
- 16-Input differential multiplexer
- Adjustable reference buffers provide 1.25, 2.048, or 2.5 V
- ADC alarm register
- Uncommitted op amps
- Dual SPDT analog switches
- Internal/external temperature sensor

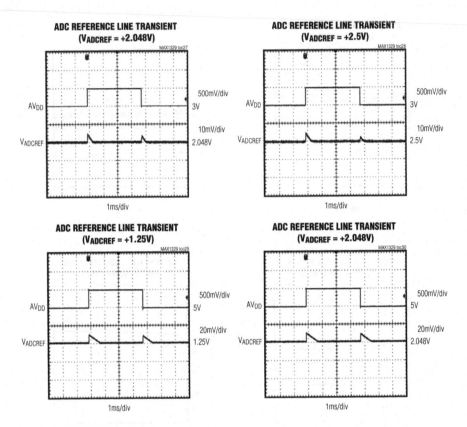

Fig. 6.3 MAX1329, typical characteristics [1]

- Internal oscillator with clock I/O (Fig. 6.6)
- Digital programmable I/O
- Analog programmable I/O
- Programmable interrupts
- Accurate supply voltage measurement
- Programmable dual voltage monitors [1]

6.1.1 Circuit Application

The MAX1329 (Figs. 6.7 and 6.8) features a 4-wire serial interface consisting of a chip select (CS), serial clock (SCLK), data in (DIN), and data out (DOUT). The data are clocked in at DIN into the shift register on the rising edge of SCLK. Data are clocked out at DOUT on the falling edge of SCLK. The serial interface is

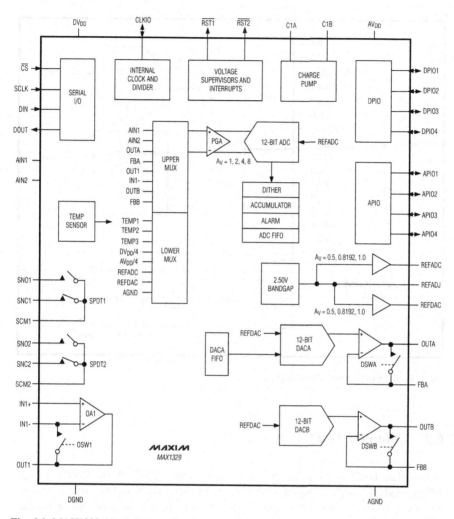

Fig. 6.4 MAX1329, block diagram [1]

compatible with SPI modes [1]. The MAX1329 performs temperature measurement by measuring the voltage across a diode-connected transistor at two different current levels. The temperature measurement process is fully automated in the MAX1329. All steps are sequenced and executed by the MAX1329 each time an input channel (or an input channel pair) configured for temperature measurement is scanned [1].

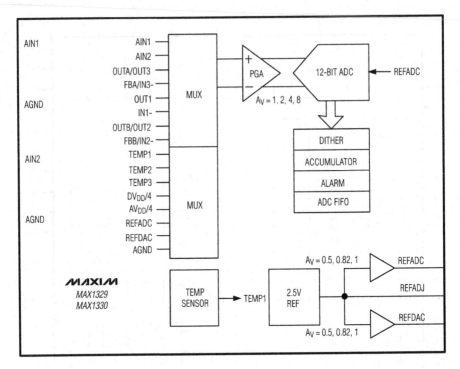

Fig. 6.5 MAX1329, block diagram [1]

6.2 Complete DAQ

A complete DAQ using MAX1329 is visualized in Fig. 6.9. A bus interface (e.g., USB) needs to manage the data with several buses. MAX3420E circuitry can be necessary to implement a full-speed USB peripheral. For microprocessor we can refer to Chap. 2. This DAQ can be used to acquire data from sensors: temperature, position sensor, and so on. In the project is included also slot expander to increase the functionality of the device: for example bluetooth module, VME module, GPS module, and so on [1–3].

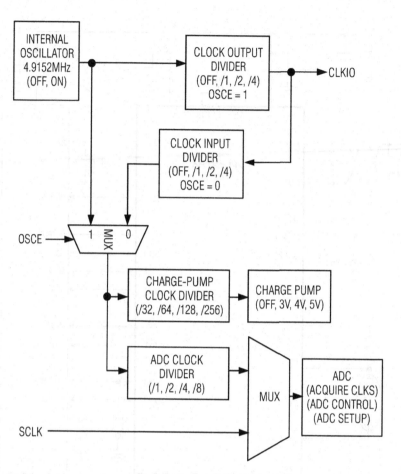

Fig. 6.6 Clock-divider block diagram [1]

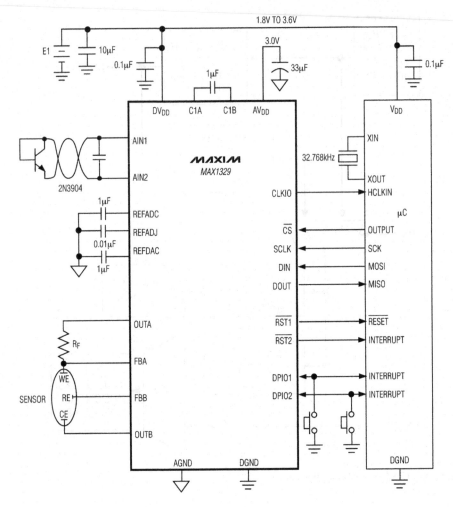

Fig. 6.7 Circuit application [1]

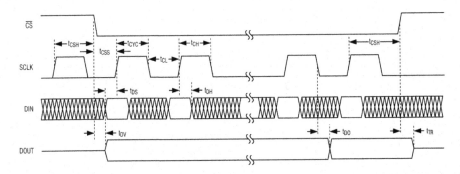

Fig. 6.8 Detailed serial-interface timing diagram [1]

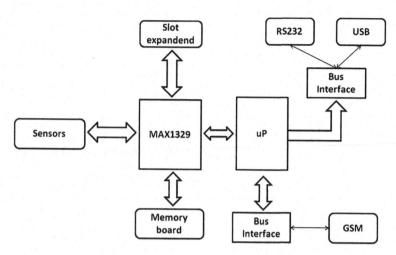

Fig. 6.9 Block diagram DAQ

References

1. Maxim Integrated (2008) Data sheet of MAX1329. Maxim Integrated. Copyright Maxim Integrated Products. http://www.maxim-ic.com. Used by permission
2. Park J, Mackqy S (2003) Practical data acquisition for instrumentation and system control. Elsevier, Oxford
3. Taylor J (1986) Computer-Based Data Acquisition System - Instrument Society of America

Index

M. Di Paolo Emilio, *Data Acquisition Systems: From Fundamentals to Applied Design*,
DOI 10.1007/978-1-4614-4214-1, © Springer Science+Business Media New York 2013

Printed in the United States
By Bookmasters